A n t h i d i u m

ns, Scop.

OSMIA

PLANORBIS

SPHEX

ARGYRONETA

tatomae

LES

CICADA

COPRIS

LAMPYRIDAE

PHAZUS

STIZUS

Epiphigger Epiphigger

The Passionate Observer

Other books by Marlene McLoughlin

The Road to Rome

Across the Aegean

Cookbooks illustrated by Marlene McLoughlin

Seafood Pasta and Noodles

Diane Seed's Rome for All Seasons

The
Passionate Observer

Writings from the world of nature by

JEAN-HENRI FABRE

Watercolors by Marlene McLoughlin

Edited by Linda Davis

CHRONICLE BOOKS

SAN FRANCISCO

ACKNOWLEDGMENTS

Our heartfelt thanks to the many people who helped us on this project. The wonderful librarians at the University of California, Berkeley, as well as our public library in Berkeley; Candace Coar and Stephanie Peters for their excellent proofreading skills and assistance in every way; George Carpenter for his support and encouragement; Alberto Rossatti for his advice and insights; and the enthusiasm of Annie Barrows and Karen Silver at Chronicle Books.

A special note of appreciation needs to be made to the translator of Fabre's writings, Teixiera de Mattos. While working on this book it has been a pleasure to sense his rapport with Fabre and feel his sensitivity and good humor towards the text.

NOTE: The insects on the title page and the page facing it are, left to right: *Podops inuncta* (Pentatoma), *Picromerus bidens* (Pentatoma), *Megalonotus chiragra*, *Graphosoma italicum* (Pentatoma), *Lygaeus saxatilis*.

Compilation copyright © 1998 by Linda Davis and Marlene McLoughlin. Illustrations copyright © 1998 by Marlene McLoughlin. All rights reserved. No part of this book may be reproduced in any form without written permission from the publisher.

Printed in Hong Kong.

ISBN 0-8118-0935-8

Library of Congress Cataloging-in-Publication Data available.

Book design: Marlene McLoughlin and Linda Davis, Star Type
Cover design: Jean Sanchirico
Composition: Star Type, Berkeley

Distributed in Canada by
Raincoast Books
8680 Cambie Street
Vancouver, B.C. V6P 6M9

10   9   8   7   6   5   4   3   2   1

Chronicle Books
85 Second Street
San Francisco, CA 94105

Web Site: www.chronbooks.com

TABLE OF CONTENTS

INTRODUCTION

What is written with pleasure is read with pleasure, according to Oscar Wilde, and it is the richness of Fabre's writing that inspired this book: the elegance, the precision, the wide range of imagery. The essays were selected from his numerous works to give the reader a sense of his life and his approach to the study of insects and how they live.

Jean Henri Fabre was born in 1823 in Saint-Léons in the south of France, west of the Rhône River, and died in Serignan, a few kilometers northwest of Orange in 1915. With the exception of a brief sojourn in Ajaccio on the island of Corsica where he taught physics and chemistry to Lycée students, he spent his life in the small area of western Provence, just north of where Cézanne and van Gogh were painting.

He taught school in Avignon (where his natural history classes at the Abbé Saint-Martial, free and open to women, scandalized the town and especially his landlady who evicted the entire Fabre family) and in Carpentras. The red ribbon of the Legion d'Honneur was awarded to him sometime around 1865. Not only did he observe with an intense gaze, he wrote with grace, corresponding and trading specimens with naturalists around the world: he also produced more than one hundred highly accomplished watercolor studies of the local fungi.

Of course I am not a scientific illustrator and my work "makes no claim to scientific responsibility," in Kenneth Clark's phrase. I've tried to give a shape to the world in which Fabre worked: the vast cobalt blue skies, the white rocky soil, the lush trees, the wind. It is now one of the prime grapegrowing areas in France, very close to Chateauneuf du Pape. I started by visiting the harmas, two rooms of which are open as a museum. One room was his laboratory, lined from floor to ceiling with glass-fronted cabinets and cases displaying a daunting number of exquisitely beautiful specimens: iridescent beetles of all sizes, spiral snail shells, bird's eggs nested in cotton, all tagged and cataloged in a minute, precise script. Then I went into the countryside to discover what I could.

Victor Hugo described Fabre as "The insects' Homer," and Darwin declared him "an incomparable observer." The combination of intense observation shaped by acute perception in writing makes Fabre a delight to read. The human world is not left behind, but extended to include these familiar but sometimes "strange" creatures.

About writing Fabre says, "My conviction is that we can say excellent things without using a barbarous vocabulary: lucidity is the sovereign politeness of the visitor. I do my best to achieve it."

Cerceris arenaria

Osma rufa

THE HARMAS

This is what I wished for, *hoc erat in votis:* a bit of land, oh, not so very large, but fenced in, to avoid the drawbacks of a public way; an abandoned, barren, sun-scorched bit of land, favored by thistles and by Wasps and Bees. Here, without fear of being troubled by the passers-by, I could consult the Ammophila and the Sphex[1] and engage in that difficult conversation whose questions and answers have experiment for their language; here, without distant expeditions that take up my time, without tiring rambles that strain my nerves, I could contrive my plans of attack, lay my ambushes and watch their effects at every hour of the day. *Hoc erat in votis.* Yes, this was my wish, my dream, always cherished, always vanishing into the mists of the future.

And it is no easy matter to acquire a laboratory in the open fields, when harassed by a terrible anxiety about one's daily bread. For forty years have I fought, with steadfast courage, against the paltry plagues of life; and the long-wished-for laboratory has come at last. What it has cost me in perseverance and relentless work I will not try to say. It has come; and, with it —

a more serious condition – perhaps a little leisure. I say perhaps, for my leg is still hampered with a few links of the convict's chain.

The wish is realized. It is a little late, O my pretty insects! I greatly fear that the peach is offered to me when I am beginning to have no teeth wherewith to eat it. Yes, it is a little late: the wide horizons of the outset have shrunk into a low and stifling canopy, more and more straitened day by day. Regretting nothing in the past, save those whom I have lost; regretting nothing, not even my first youth; hoping nothing either, I have reached the point at which, worn out by the experience of things, we ask ourselves if life be worth the living.

Amid the ruins that surround me, one strip of wall remains standing, immovable upon its solid base: my passion for scientific truth. Is that enough, O my busy insects, to enable me to add yet a few seemly pages to your history? Will my strength not cheat my good intentions? Why, indeed, did I forsake you so long? Friends have reproached me for it. Ah, tell them, tell those friends, who are yours as well as mine, tell them that it was not forgetfulness on my part, not weariness, nor neglect: I thought of you; I was convinced that the Cerceris[2] cave had more fair secrets to reveal to us, that the chase of the Sphex held fresh surprises in store. But time failed me; I was alone, deserted, struggling against misfortune. Before philosophizing, one had to live. Tell them that; and they will pardon me.

Others again have reproached me with my style, which has not the solemnity, nay, better, the dryness of the schools. They fear lest a page that is read without fatigue should not always be the expression of the truth. Were I to take their word for it, we are profound only on condition of being obscure. Come here, one and all of you – you, the sting-bearers, and you, the wing-cased armour-clads – take up my defense and bear witness in my favor. Tell of the intimate terms on which I live with you, of the patience with which I observe you, of the care with which I record your actions. Your evidence is unanimous: yes, my pages, though they bristle not with hollow formulas nor learned smatterings, are the exact narrative of facts observed, neither more nor less; and whoso cares to question you in his turn will obtain the same replies.

And then, my dear insects, if you cannot convince those good people, because you do not carry the weight of tedium, I, in my turn, will say to them:

You rip up the animal and I study it alive; you turn it into an object of horror and pity, whereas I cause it to be loved; you labor in a torture-chamber and dissecting-room, I make my observations under the blue sky to the song of the Cicadas,[3] you subject cell and protoplasm to chemical tests, I study instinct in its loftiest manifestations; you pry into death, I pry into life. And why should I not complete my thought: the boars have muddied the clear stream; natural history, youth's glorious study, has, by dint of cellular improvements, become a hateful and repulsive thing. Well, if I write for men of learning, for philosophers, who, one day, will try to some extent to unravel the tough problem of instinct, I write also, I write above all things for the young. I want to make them love the natural history which you make them hate; and that is why, while keeping strictly to the domain of truth, I avoid your scientific prose, which too often, alas seems borrowed from some Iroquois idiom!

But this is not my business for the moment: I want to speak of the bit of land long cherished in my plans to form a laboratory of living entomology, the bit of land which I have at last obtained in the solitude of a little village. It is a *harmas,* the name given, in this district,[4] to an untilled, pebbly expanse abandoned to the vegetation of the thyme. It is too poor to repay the work of the plough; but the sheep passes there in spring, when it has chanced to rain and a little grass shoots up.

My harmas, however, because of its modicum of red earth swamped by a huge mass of stones, has received a rough first attempt at cultivation: I am told that vines once grew here. And, in fact, when we dig the ground before planting a few trees, we turn up, here and there, remains of the precious stock, half-carbonized by time. The three-pronged fork, therefore, the only implement of husbandry that can penetrate such a soil as this, has entered here; and I am sorry, for the primitive vegetation has disappeared. No more thyme, no more lavender, no more clumps of kermes-oak, the dwarf oak that forms forests across which we step by lengthening our stride a little. As these plants, especially the first two, might be of use to me by offering the Bees and Wasps a spoil to forage, I am compelled to reinstate them in the ground whence they were driven by the fork.

What abounds without my mediation is the invaders of any soil that is first dug up and then left for a long time to its own resources. We have, in the first rank, the couch-grass, that execrable weed which three years of stubborn warfare have not succeeded in exterminating. Next, in respect of number, come the centauries, grim-looking one and all, bristling with prickles or starry halberds. They are the yellow-flowered centaury, the mountain centaury, the star-thistle and the rough centaury: the first predominates. Here and there, amid their inextricable confusion, stands, like a chandelier with spreading, orange flowers for lights, the fierce Spanish oyster-plant, whose spikes are strong as nails. Above it, towers the Illyrian cotton-thistle, whose straight and solitary stalk soars to a height of three

Andrena armata

Centaurea solstitialis

to six feet and ends in large pink tufts. Its armor hardly yields before that of the oyster-plant. Nor must we forget the lesser thistle-tribe, with first of all, the prickly or "cruel" thistle, which is so well armed that the plant-collector knows not where to grasp it; next, the spear-thistle, with its ample foliage, ending each of its veins with a spear-head; lastly, the black knap-weed, which gathers itself into a spiky knot. In among these, in long lines armed with hooks, the shoots of the blue dewberry creep along the ground. To visit the prickly thicket when the Wasp goes foraging, you must wear boots that come to mid-leg or else resign yourself to a smarting in the calves. As long as the ground retains a few remnants of the vernal rains, this rude vegetation does not lack a certain charm, when the pyramids of the oyster-plant and the slender branches of the cotton-thistle rise above the wide carpet formed by the yellow-flowered centaury saffron heads; but let the droughts of summer come and we see but a desolate waste, which the flame of a match would set ablaze from one end to the other. Such is, or rather was, when I took possession of it, the Eden of bliss where I mean to live henceforth with the insect. Forty years of desperate struggle have won it for me.

Eden, I said; and, from the point of view that interests me, the expression is not out of place. This cursed ground, which no one would have had at a gift to sow with a pinch of turnip-seed, is an earthly paradise for the Bees and Wasps. Its mighty growth of thistles and centauries draws them all to me from everywhere around. Never, in my insect-hunting memories, have I seen so large a population at a single spot; all the trades have made it their rallying-point. Here come hunters of every kind of game, builders in clay, weavers of cotton goods, collectors of pieces cut from a leaf or the petals of a flower, architects in pasteboard, plasterers mixing mortar, carpenters boring wood, miners digging underground galleries, workers handling goldbeater's skin and many more.

Who is this one? An Anthidium.[5] She scrapes the cobwebby stalk of the yellow-flowered centaury and gathers a ball of wadding which she carries

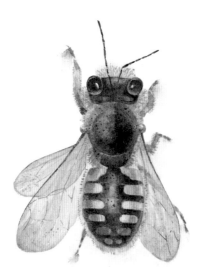

Anthidium florentium

off proudly in the tips of her mandibles. She will turn it, under ground, into cotton-felt satchels to hold the store of honey and the egg. And these others, so eager for plunder? They are Megachiles,[6] carrying under their bellies their black, white or blood-red reaping-brushes. They will leave the thistles to visit the neighboring shrubs and there cut from the leaves oval pieces which will be made into a fit receptacle to contain the harvest. And these, clad in black velvet? They are Chalicodomæ,[7] who work with cement and gravel. We could easily find their masonry on the stones in the harmas. And these, noisily buzzing with a sudden flight? They are the Anthophoræ,[8] who live in the old walls and the sunny banks of the neighborhood.

Megachile centuncularis

Now come the Osmiæ. One stacks her cells in the spiral staircase of an empty snail-shell; another, attacking the pith of a dry bit of bramble, obtains for her grubs a cylindrical lodging and divides it into floors by means of partition-walls; a third employs the natural channel of a cut reed; a fourth is a rent-free tenant of the vacant galleries of some Mason-bee. Here are the Macroceræ and the Euceræ, whose males are proudly horned; the Dasypodæ, who carry an ample brush of bristles on their hind-legs for a reaping implement; the Andrenæ, so manifold in species; the slender-bellied Halicti.[9] I omit a host of others. If I tried to continue this record of the guests of my thistles, it would muster almost the whole of the honey-yielding tribe. A learned entomologist of Bordeaux, Professor Pérez, to whom I submit the naming of my prizes, once asked me if I had any special means of hunting, to send him so many rarities and even novelties. I am not at all an experienced and, still less, a zealous hunter, for the insect interests me much more when engaged in its work than when stuck on a pin in a cabinet. The whole secret of my hunting is reduced to my dense nursery of thistles and centauries.

Dasypoda hirtipes

By a most fortunate chance, with this populous family of honey-gatherers was allied the whole hunting tribe. The builders'

9

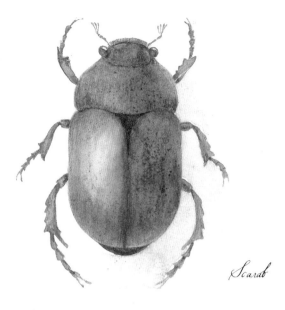

Scarab

men had distributed here and there in the harmas great mounds of sand and heaps of stones, with a view to running up some surrounding walls. The work dragged on slowly; and the materials found occupants from the first year. The Mason-bees had chosen the interstices between the stones as a dormitory where to pass the night, in serried groups. The powerful Eyed Lizard, who, when close-pressed, attacks both man and dog, wide-mouthed, had selected a cave wherein to lie in wait for the passing Scarab;[10] the Black-eared Chat, garbed like a Dominican, white-frocked with black wings, sat on the top stone, singing his short rustic lay: his nest, with its sky-blue eggs, must be somewhere in the heap. The little Dominican disappeared with the loads of stones. I regret him: he would have been a charming neighbor. The Eyed Lizard I do not regret at all.

The sand sheltered a different colony. Here, the Bembeces[11] were sweeping the threshold of their burrows, flinging a curve of dust behind them; the Languedocian Sphex was dragging her Ephippigera[12] by the antennæ; a Stizus[13] was storing her preserves of Cicadellæ.[14] To my sorrow, the

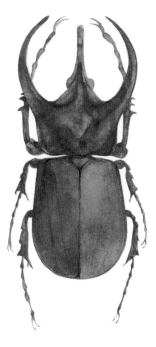

Lamellicornia

masons ended by evicting the sporting tribe; but, should I ever wish to recall it, I have but to renew the mounds of sand: they will soon all be there.

Hunters that have not disappeared, their homes being different, are the Ammophilæ, whom I see fluttering, one in spring, the others in autumn, along the garden-walks and over the lawns, in search of a Caterpillar; the Pompili,[15] who travel alertly, beating their wings and rummaging in every corner in quest of a Spider. The largest of them waylays the Narbonne Lycosa,[16] whose burrow is not infrequent in the harmas. This burrow is a vertical well, with a curb of fescue-grass intertwined with silk. You can see the eyes of the mighty Spider gleam at the bottom of the den like little diamonds, an object of terror to most. What a prey and what dangerous hunting for the Pompilus! And here, on a hot summer afternoon, is the Amazon-ant, who leaves her barrack-rooms in long battalions and marches far afield to hunt for slaves. We will follow her in her raids when we find time. Here again, around a heap of grasses turned to mold, are Scoliæ[17] an inch and a half long, who fly gracefully and dive into the heap, attracted by a rich prey, the grubs of Lamellicorns, Oryctes and Cetoniæ.[18]

What subjects for study! And there are more to come. The house was as utterly deserted as the ground. When man was gone and peace assured, the animal hastily seized on everything. The Warbler took up his abode in the lilac-shrubs; the Greenfinch settled in the thick shelter of the cypresses; the Sparrow carted rags and straw under every slate; the Serin-finch, whose downy nest is no bigger than half an apricot, came and chirped in the plane-tree-tops; the Scops made a habit of uttering his monotonous, piping note here, of an evening; the bird of Pallas Athene, the Owl, came hurrying along to hoot and hiss.

In front of the house is a large pond, fed by the aqueduct that supplies the village-pumps with water. Here, from half a mile and more around, come the Frogs and Toads in the lovers' season. The Natterjack, sometimes as large as a plate, with a narrow stripe of yellow down his back, makes his appointments here to take his bath; when the evening twilight falls, we see hopping along the edge the Midwife Toad, the male, who carries a cluster of eggs, the size of peppercorns, wrapped around his hind-legs: the genial paterfamilias has brought his precious packet from afar, to leave it in the water and afterwards retire under some flat stone, whence he will emit a sound like a tinkling bell. Lastly, when not croaking amid the foliage, the Tree-frogs indulge in the most graceful dives. And so, in May, as soon as it is dark, the pond becomes a deafening orchestra: it is impossible to talk at table, impossible to sleep. We had to remedy this by means perhaps a little too rigorous. What could we do? He who tries to sleep and cannot needs becomes ruthless.

Bolder still, the Wasp has taken possession of the dwelling-house. On my door-sill, in a soil of rubbish, nestles the White-banded Sphex: when I go indoors, I must be careful not to damage her burrows, not to tread upon the miner absorbed in her work. It is quite a quarter of a century since I last saw the saucy Cricket-hunter. When I made her acquaintance, I used to visit her at a few miles' distance: each time, it meant an expedition under the blazing August sun. Today, I find her at my door; we are intimate neighbors. The embrasure of the closed window provides an apartment of a mild temperature for the Pelopæus.[19] The earth-built nest is fixed against the freestone wall. To enter her home, the Spider-huntress uses a little hole left open by accident in the shutters. On the moldings of the Venetian blinds, a few stray Mason-bees build their group of cells; inside the outer shutters, left ajar, a Eumenes[20] constructs her little earthen dome, surmounted by a short, bell-mouthed neck. The common Wasp and the Polistes[21] are my dinner-guests: they visit my table to see if the grapes served are as ripe as they look.

Here, surely – and the list is far from complete – is a company both numerous and select, whose conversation will not fail to charm my solitude, if I succeed in drawing it out. My dear beasts of former days, my old friends, and others, more recent acquaintances, all are here, hunting, foraging, building in close proximity. Besides, should we wish to vary the scene of observation, the mountain[22] is but a few hundred steps away, with its tangle of arbutus, rock-roses and arborescent heather; with its sandy spaces dear to the Bembeces; with its marly slopes exploited by different Wasps and Bees. And that is why, foreseeing these riches, I have abandoned the town for the village and come to Sérignan to weed my turnips and water my lettuces.

Laboratories are being founded, at great expense, on our Atlantic and Mediterranean coasts, where people cut up small sea-animals, of but meager interest to us; they spend a fortune on powerful microscopes, delicate dissecting-instruments, engines of capture, boats, fishing-crews, aquariums, to find out how the yolk of an Annelid's[23] egg is constructed, a question whereof I have never yet been able to grasp the full importance; and they scorn the little land-animal, which lives in constant touch with us,

Anthophora acervorum

Anthophora hispanica

which provides universal psychology with documents of inestimable value, which too often threatens the public wealth by destroying our crops. When shall we have an entomological laboratory for the study not of the dead insect, steeped in alcohol, but of the living insect; a laboratory having for its object the instinct, the habits, the manner of living, the work, the struggles, the propagation of that little world, with which agriculture and philosophy have most seriously to reckon?

To know thoroughly the history of the destroyer of our vines might perhaps be more important than to know how this or that nerve-fiber of a Cirriped[24] ends; to establish by experiment the line of demarcation between intellect and instinct; to prove, by comparing facts in the zoological progression, whether human reason be an irreducible faculty or not: all this ought surely to take precedence of the number of joints in a Crustacean's antenna. These enormous questions would need an army of workers; and we have not one. The fashion is all for the Mollusc and the Zoophytes.[25] The depths of the sea are explored with many drag-nets; the soil which we tread is consistently disregarded. While waiting for the fashion to change, I open my harmas laboratory of living entomology; and this laboratory shall not cost the ratepayers one farthing.

NOTES

1. Two species of Digger or Hunting Wasps. See *Insect Life,* by J. H. Fabre, translated by the author of *Mademoiselle Mori:* chaps. vi to xii and xvi. – *Translator's Note.*

2. A species of Digger Wasp. See *Insect Life:* chaps. vi to xii and xvi. – *Translator's Note.*

3. The Cicada is the *Cigale,* an insect akin to the Grasshopper and found more particularly in the south of France. See *Social Life in the Insect World,* by J. H. Fabre, translated by Bernard Miall: chaps. i to iv. – *Translator's Note.*

4. The country around Sérignan, in Provence. – *Translator's Note.*

5. A Tailor-bee. – *Translator's Note.*

6. Leaf-cutting Bees. – *Translator's Note.*

7. Mason-bees. See *Insect Life:* chaps. xx to xxii. – *Translator's Note.*

8. A species of Wild Bees. – *Translator's Note.*

9. Osmiæ, Macroceræ, Euceræ, Dasypodæ, Andrenæ and Haliciti are all different species of Wild Bees. For the Haliciti, see *The Life and Love of the Insect,* by J. Henri Fabre, translated by Alexander Teixeira de Mattos: chaps. xv and xvi. – *Translator's Note.*

10. A Dung-beetle also known as the Sacred Beetle. See *Insect Life:* chaps. i and ii; and *The Life and Love of the Insect:* chaps. i to iv. – *Translator's Note.*

11. A species of Digger-wasps. See *Insect Life:* chap. xvi. – *Translator's Note.*

12. A species of Green Grasshopper. – *Translator's Note.*

13. A species of Hunting Wasp. – *Translator's Note.*

14. Froghoppers. – *Translator's Note.*

15. The Pompilus is a species of Digger or Hunting Wasp, known also as the Ringed Calicurgus. See *The Life and Love of the Insect:* chap. xii. – *Translator's Note.*

16. Known also as the Black-bellied Tarantula. See *The Life and Love of the Insect:* chap. xii; and *The Life of the Spider:* chaps. i and iii to vi. – *Translator's Note.*

17. Large Hunting Wasps. See *The Life and Love of the Insect:* chap. xi. – *Translator's Note.*

18. Different species of Beetles. The Cetonia is the Rose-chafer. – *Translator's Note.*

19. A species of Mason-wasp. – *Translator's Note.*

20. A species of Mason-wasp. – *Translator's Note.*

21. A species of Solitary Wasp. – *Translator's Note.*

22. Mont Ventoux, an outlying summit of the Alps, 6,270 feet high. See *Insect Life:* chap. xiii. – *Translator's Note.*

23. A red-blooded Worm. – *Translator's Note.*

24. Cirripeds are sea-animals with hair-like legs, including the Barnacles and Acorn-shells. – *Translator's Note.*

25. Zoophytes are plant-like sea-animals, including Star-fishes, Jelly-fishes, Sea-anemones and Sponges. – *Translator's Note.*

Anthophora
acervorum

Tettigonia viridissima

THE GREEN GRASSHOPPER

We are in the middle of July. The astronomical dog-days are just beginning; but in reality the torrid season has anticipated the calendar and for
some weeks past the heat has been overpowering.

This evening in the village they are celebrating the National Festival.[1]
While the little boys and girls are hopping around a bonfire whose gleams
are reflected upon the church-steeple, while the drum is pounded to mark
the ascent of each rocket, I am sitting alone in a dark corner, in the comparative coolness that prevails at nine o'clock, harking to the concert of
the festival of the fields, the festival of the harvest, grander by far than
that which, at this moment, is being celebrated in the village square with
gunpowder, lighted torches, Chinese lanterns and, above all, strong drink.
It has the simplicity of beauty and the repose of strength.

It is late; and the Cicadæ are silent. Glutted with light and heat, they
have indulged in symphonies all the livelong day. The advent of the night
means rest for them, but a rest frequently disturbed. In the dense branches
of the plane-trees, a sudden sound rings out like a cry of anguish, strident
and short. It is the desperate wail of the Cicada, surprised in his quietude
by the Green Grasshopper, that ardent nocturnal huntress, who springs

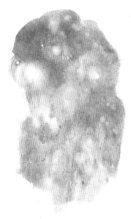

upon him, grips him in the side, opens and ransacks his abdomen. An orgy of music, followed by butchery.

I have never seen and never shall see that supreme expression of our national revelry, the military review at Longchamp; nor do I much regret it. The newspapers tell me as much about it as I want to know. They give me a sketch of the site. I see, installed here and there amid the trees, the ominous Red Cross, with the legend, "Military Ambulance; Civil Ambulance." There will be bones broken, apparently; cases of sunstroke; regrettable deaths, perhaps. It is all provided for and all in the program.

Even here, in my village, usually so peaceable, the festival will not end, I am ready to wager, without the exchange of a few blows, that compulsory seasoning of a day of merry-making. No pleasure, it appears, can be fully relished without an added condiment of pain.

Let us listen and meditate far from the tumult. While the disembowelled Cicada utters his protest, the festival up there in the plane-trees is continued with a change of orchestra. It is now the time of the nocturnal performers. Hard by the place of slaughter, in the green bushes, a delicate ear perceives the hum of the Grasshoppers. It is the sort of noise that a spinning-wheel makes, a very unobtrusive sound, a vague rustle of dry membranes rubbed together. Above this dull bass there rises, at intervals, a hurried, very shrill, almost metallic clicking. There you have the air and the recitative, intersected by pauses. The rest is the accompaniment.

Despite the assistance of a bass, it is a poor concert, very poor indeed, though there are about ten executants in my immediate vicinity. The tone lacks intensity. My old tympanum is not always capable of perceiving these subtleties of sound. The little that reaches me is extremely sweet and most appropriate to the calm of twilight. Just a little more breadth in your bow-stroke, my dear Green Grasshopper, and your technique would be better

than the hoarse Cicada's, whose name and reputation you have been made to usurp in the countries of the north.

Still, you will never equal your neighbor, the little bell-ringing Toad, who goes tinkling all round, at the foot of the plane-trees, while you click up above. He is the smallest of my batrachian folk and the most venturesome in his expeditions.

How often, at nightfall, by the last glimmers of daylight, have I not come upon him as I wandered through my garden, hunting for ideas! Something runs away, rolling over and over in front of me. Is it a dead leaf blown along by the wind? No, it is the pretty little Toad disturbed in the midst of his pilgrimage. He hurriedly takes shelter under a stone, a clod of earth, a tuft of grass, recovers from his excitement and loses no time in picking up his liquid note.

On this evening of national rejoicing, there are nearly a dozen of him tinkling against one another around me. Most of them are crouching among the rows of flower-pots that form a sort of lobby outside my house. Each has his own note, always the same, lower in one case, higher in another, a short, clear note, melodious and of exquisite purity.

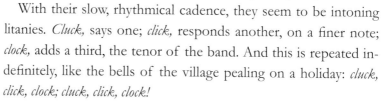

With their slow, rhythmical cadence, they seem to be intoning litanies. *Cluck,* says one; *click,* responds another, on a finer note; *clock,* adds a third, the tenor of the band. And this is repeated indefinitely, like the bells of the village pealing on a holiday: *cluck, click, clock; cluck, click, clock!*

The batrachian choristers remind me of a certain harmonica which I used to covet when my six-year-old ear began to awaken to the magic of sounds. It consisted of a series of strips of glass of unequal length, hung on two stretched tapes. A cork fixed to a wire served as a hammer. Imagine an unskilled hand striking at random on this key-board, with a sudden clash of octaves, dissonances and topsy-turvy chords; and you will have a pretty clear idea of the Toads' litany.

As a song, this litany has neither head nor tail to it; as a collection of pure sounds, it is delicious. This is the case with all the music in nature's concerts. Our ear discovers superb notes in it and then becomes refined and acquires, outside the realities of sound, that sense of order which is the first condition of beauty.

Now this sweet ringing of bells between hiding-place and hiding-place is the matrimonial oratorio, the discreet summons which every Jack issues to his Jill. The sequel to the concert may be guessed without further enquiry; but what it would be impossible to foresee is the strange finale of the wedding. Behold the father, in this case a real *paterfamilias,* in the noblest sense of the word, coming out of his retreat one day in an unrecognizable state. He is carrying the future, tightpacked around his hind-legs; he is changing houses laden with a cluster of eggs the size of pepper-corns. His calves are girt, his thighs are sheathed with the bulky burden; and it covers his back like a beggar's wallet, completely deforming him.

Whither is he going, dragging himself along, incapable of jumping, thanks to the weight of his load? He is going, the fond parent, where the mother refuses to go; he is on his way to the nearest pond, whose warm waters are indispensable to the tadpoles' hatching and existence. When the eggs are nicely ripened around his legs under the humid shelter of a stone, he braves the damp and the daylight, he the passionate lover of dry land and darkness; he advances by short stages, his lungs congested with fatigue. The pond is far away, perhaps; no matter: the plucky pilgrim will find it.

He's there. Without delay, he dives, despite his profound antipathy to bathing; and the cluster of eggs is instantly removed by the legs rubbing against each other. The eggs are now in their element; and the rest will be accomplished of itself. Having fulfilled his obligation to go right under, the father hastens to return to his well-sheltered home. He is scarcely out of sight before the little black tadpoles are hatched and playing about. They were but waiting for the contact of the water in order to burst their shells.

Among the singers in the July gloaming, one alone, were he able to vary his notes, could vie with the Toad's harmonious bells. This is the little Scops Owl, that comely nocturnal bird of prey, with the round gold eyes. He sports on his forehead two small feathered horns which have won for him in the district the name of *Machoto banarudo,* the Horned Owl. His song, which is rich enough to fill by itself the still night air, is of a nerve-shattering monotony. With imperturbable and measured regularity, for hours on end, *kew, kew,* the bird spits out its cantata to the moon.

Scops owl

One of them has arrived at this moment, driven from the plane-trees in the square by the din of the rejoicings, to demand my hospitality. I can hear him in the top of a cypress near by. From up there, dominating the lyrical assembly, at regular intervals he cuts into the vague orchestration of the Grasshoppers and the Toads.

His soft note is contrasted, intermittently, with a sort of Cat's mew, coming from another spot. This is the call of the Common Owl, the meditative bird of Minerva. After hiding all day in the seclusion of a hollow olive-tree, he started on his wanderings when the shades of evening began to fall. Swinging along with a sinuous flight, he came from somewhere in the neighborhood to the pines in my enclosure, whence he mingles his harsh mewing, slightly softened by distance, with the general concert.

The Green Grasshopper's clicking is too faint to be clearly perceived amidst these clamorers; all that reaches me is the least ripple, just noticeable when there is a moment's silence. He possesses as his apparatus of sound only a modest drum and scraper, whereas they, more highly privileged, have their bellows, the lungs, which send forth a column of vibrating air. There is no comparison possible. Let us return to the insects.

One of these, though inferior in size and no less sparingly equipped, greatly surpasses the Grasshopper in nocturnal rhapsodies. I speak of the pale and slender Italian Cricket (*Œcanthus pellucens,* Scop.), who is so puny that you dare not take him up for fear of crushing him. He makes music everywhere among the rosemary-bushes, while the Glow-worms light up their blue lamps to complete the revels. The delicate instrumentalist consists chiefly of a pair of large wings, thin and gleaming as strips of mica. Thanks to these dry sails, he fiddles away with an intensity capable of drowning the Toads' fugue. His performance suggests, but with more brilliancy, more *tremolo* in the execution, the song of the Common Black Cricket. Indeed the mistake would certainly be made by any one who did not know that, by the time that the very hot weather comes, the true Cricket, the chorister of spring, has disappeared. His pleasant violin has been succeeded by another more pleasant still and worthy of special study. We shall return to him at an opportune moment.

These then, limiting ourselves to select specimens, are the principal participants in this musical evening: the Scops Owl, with his languorous solos;

Lampyris noctiluca

the Toad, that tinkler of sonatas; the Italian Cricket, who scrapes the first string of a violin; and the Green Grasshopper, who seems to beat a tiny steel triangle.

We are celebrating to-day, with greater uproar than conviction, the new era, dating politically from the fall of the Bastille; they, with glorious indifference to human things, are celebrating the festival of the sun, singing the happiness of existence, sounding the loud hosanna of the July heats.

What care they for man and his fickle rejoicings! For whom or for what will our squibs be spluttering a few years hence? Far-seeing indeed would he be who could answer the question. Fashions change and bring us the unexpected. The time-serving rocket spreads its sheaf of sparks for the public enemy of yesterday, who has become the idol of to-day. Tomorrow it will go up for somebody else.

In a century or two, will any one, outside the historians, give a thought to the taking of the Bastille? It is very doubtful. We shall have other joys and also other cares.

Let us look a little farther ahead. A day will come, so everything seems to tell us, when, after making progress upon progress, man will succumb, destroyed by the excess of what he calls civilization. Too eager to play the god, he cannot hope for the animal's placid longevity; he will have disappeared when the little Toad is still saying his litany, in company with the Grasshopper, the Scops Owl and the others. They were singing on this planet before us; they will sing after us, celebrating what can never change, the fiery glory of the sun.

I will dwell no longer on this festival and will become once more the naturalist, anxious to obtain information concerning the private life of the insect. The Green Grasshopper (*Locusta viridissima,* Lin.) does not appear to be common in my neighborhood. Last year, intending to make a study of this insect and finding my efforts to hunt it fruitless, I was obliged to have recourse to the good offices of a forest-ranger, who sent me a pair of couples from the Lagarde plateau, that bleak district where the beech-tree begins its escalade of the Ventoux.

Now and then freakish fortune takes it into her head to smile upon the persevering. What was not to be found last year has become almost common this summer. Without leaving my narrow enclosure, I obtain as many Grasshoppers as I could wish. I hear them rustling at night in the green thickets. Let us make the most of the windfall, which perhaps will not occur again.

In the month of June, my treasures are installed, in a sufficient number of couples, under a wire cover standing on a bed of sand in an earthen pan. It is indeed a magnificent insect, pale-green all over, with two whitish stripes running down its sides. Its imposing size, its slim proportions and its great gauze wings make it the most elegant of our Locustidæ. I am enraptured with my captives. What will they teach me? We shall see. For the moment, we must feed them.

I have here the same difficulty that I had with the Decticus. Influenced

by the general diet of the Orthoptera,[2] those ruminants of the green-swards, I offer the prisoners a leaf of lettuce. They bite into it, certainly, but very sparingly and with a scornful tooth. It soon becomes plain that I am dealing with half-hearted vegetarians. They want something else: they are beasts of prey, apparently. But what manner of prey? A lucky chance taught me.

At break of day I was pacing up and down outside my door, when something fell from the nearest plane-tree with a shrill grating sound. I ran up and saw a Grasshopper gutting the belly of an exhausted Cicada. In vain the victim buzzed and waved his limbs: the other did not let go, dipping her head right into the entrails and rooting them out by small mouthfuls.

I knew what I wanted to know: the attack had taken place up above, early in the morning, while the Cicada was asleep; and the plunging of the poor wretch, dissected alive, had made assailant and assailed fall in a bundle to the ground. Since then I have repeatedly had occasion to witness similar carnage.

I have even seen the Grasshopper – the height of audacity, this – dart in pursuit of a Cicada in mad flight. Even so does the Sparrow Hawk pursue the Swallow in the sky. But the bird of prey here is inferior to the insect. It attacks one weaker than itself. The Grasshopper, on the other hand, assaults a colossus, much larger than herself and stronger; and nevertheless the result of the unequal fight is not in doubt. The Grasshopper rarely fails with the sharp pliers of her powerful jaws to disembowel her capture, which, being unprovided with weapons, confines itself to crying out and kicking.

The main thing is to retain one's hold of the prize, which is not difficult in somnolent darkness. Any Cicada encountered by the fierce Locustid on her nocturnal rounds is bound to die a lamentable death. This explains those sudden agonized notes which grate through the woods at late, unseasonable hours, when the cymbals have long been silent. The murderess in her suit of apple-green has pounced on some sleeping Cicada.

My boarders' menu is settled: I will feed them on Cicadæ. They take such a liking to this fare that, in two or three weeks, the floor of the cage is a knacker's yard strewn with heads and empty thoraces, with torn-off wings and disjointed legs. The belly alone disappears almost entirely. This is the tit-bit, not very substantial, but extremely tasty, it would seem. Here, in fact, in the insect's crop, the syrup is accumulated, the sugary sap which the Cicada's gimlet taps from the tender bark. Is it because of this dainty that the prey's abdomen is preferred to any other morsel? It is quite possible.

I do, in fact, with a view to varying the diet, decide to serve up some very sweet fruits, slices of pear, grape-pips, bits of melon. All this meets with delighted appreciation. The Green Grasshopper resembles the English: she dotes on underdone rump-steak seasoned with jam.[3] This perhaps is why, on catching the Cicada, she first rips up his paunch, which supplies a mixture of flesh and preserves.

To eat Cicadæ and sugar is not possible in every part of the country. In the north, where she abounds, the Green Grasshopper would not find the dish which attracts her so strongly here. She must have other resources. To convince myself of this, I give her Anoxiæ (*A. pilosa,* Fab.), the summer equivalent of the spring Cockchafer. The Beetle is accepted without hesitation. Nothing is left of him but the wing-cases, head and legs. The result is the same with the magnificent plump Pine Cockchafer (*Melolontha fullo,* Lin.), a sumptuous morsel which I find next day eviscerated by my gang of knackers.

These examples teach us enough. They tell us that the Grasshopper is an inveterate consumer of insects, especially of those which are not protected by too hard a cuirass; they are evidence of tastes which are highly carnivorous, but not exclusively so, like those of the Praying Mantis, who refuses everything except game. The butcher of the Cicadæ is able to modify an excessively heating diet with vegetable fare. After meat and blood, sugary fruit-pulp; sometimes even, for lack of anything better, a little green stuff.

Neverthess, cannibalism is prevalent. True, I never witness in my Grasshopper cages the savagery which is so common in the Praying Mantis, who harpoons her rivals and devours her lovers; but, if some weakling succumb, the survivors hardly ever fail to profit by his carcass as they would in the case of any ordinary prey. With no scarcity of provisions as an excuse, they feast upon their defunct companion. For the rest, all the saber-bearing clan display, in varying degrees, a propensity for filling their bellies with their maimed comrades.

In other respects, the Grasshoppers live together very peacefully in my cages. No serious strife ever takes place among them, nothing beyond a little rivalry in the matter of food. I hand in a piece of pear. A Grasshopper alights on it at once. Jealously she kicks away any one trying to bite at the delicious morsel. Selfishness reigns everywhere. When she has eaten her fill, she makes way for another, who in her turn becomes intolerant. One after the other, all the inmates of the menagerie come and refresh themselves. After cramming their crops, they scratch the soles of their feet a little with their mandibles, polish up their forehead and eyes with a leg moistened with spittle and then, hanging to the trelliswork or lying on the sand in a posture of contemplation, blissfully they digest and slumber most of the day, especially during the hottest part of it.

It is in the evening, after sunset, that the troop becomes lively. By nine o'clock the animation is at its height. With sudden rushes they clamber to the top of the dome, to descend as hurriedly and climb up once more. They come and go tumultuously, run and hop around the circular track and, without stopping, nibble at the good things on the way.

The males are stridulating by themselves, here and there, teasing the passing fair with their antennæ. The future mothers stroll about gravely, with their saber half-raised. The agitation and feverish excitement means that the great business of pairing is at hand. The fact will escape no practiced eye.

It is also what I particularly wish to observe. My chief object in stocking my cages was to discover how far the strange nuptial manners revealed by the White-faced Decticus might be regarded as general. My wish is satisfied, but not fully, for the late hours at which events take place did not allow me to witness the final act of the wedding. It is late at night or early in the morning that things happen.

The little that I see is confined to interminable preludes. Standing face to face, with foreheads almost touching, the lovers feel and sound each other for a long time with their limp antennæ. They suggest two fencers crossing and recrossing harmless foils. From time to time, the male stridulates a little, gives a few short strokes of the bow and then falls silent, feeling perhaps too much overcome to continue. Eleven o'clock strikes; and the declaration is not yet over. Very regretfully, but conquered by sleepiness, I quit the couple.

Next morning, early, the female carries, hanging at the bottom of her ovipositor, the queer bladderlike arrangement that surprised us so much in the Decticus. It is an opaline capsule, the size of a large pea and roughly subdivided into a small number of egg-shaped vesicles. When the Grasshopper walks, the thing scrapes along the ground and becomes dirty with sticky grains of sand.

The final banquet of the female Decticus is seen again here in all its hideousness. When, after a couple of hours, the fertilizing capsule is drained of its contents, the Grasshopper devours it bit by bit; for a long time she chews and rechews the gummy morsel and ends by swallowing it all down. In less than half a day, the milky burden has disappeared, consumed with zest down to the last atom.

The inconceivable therefore, imported, one would think, from another planet, so far removed is it from earthly habits, reappears with no noticeable variation in the Grasshopper, following on the Decticus. What singular folk are the Locustidæ, one of the oldest races in the animal kingdom on dry land! It seems probable that these eccentricities are the rule throughout the order. Let us consult another saber-bearer.

I select the Ephippiger (*Ephippigera vitium*, Serv.), who is so easy to rear on bits of pear and lettuce-leaves. It is in July and August that things happen. A little way off, the male is stridulating by himself. His ardent bow-strokes set his whole body quivering. Then he stops. Little by little, with slow and almost ceremonious steps, the caller and the called come closer together. They stand face to face, both silent, both stationary, their antennæ gently swaying, their fore-legs raised awkwardly and giving a sort of

handshake at intervals. The peaceful interview lasts for hours. What do they say to each other? What vows do they exchange? What does their ogling mean?

But the moment has not yet come. They separate, they fall out and each goes his own way. The coolness does not last long. Here they are together again. The tender declarations are resumed, with no more success than before. At last, on the third day, I behold the end of the preliminaries. The male slips discreetly under his companion, backwards, according to the immemorial laws and customs of the Crickets. Stretched out behind and lying on his back, he clings to the ovipositor, his prop. The pairing is accomplished.

The result is an enormous spermatophore, a sort of opalescent raspberry with large seeds. Its color and shape remind one of a cluster of Snail's-eggs. I remember seeing the same effect once with a Decticus, but in a less striking form; and I find it again in the Green Grasshopper's spermatophore. A thin median groove divides the whole into two symmetrical bunches, each comprising seven or eight spherules. The two nodes situated right and left of the bottom of the ovipositor are more transparent than the others and contain a bright orange-red kernel. The whole thing is attached by a wide pedicle, a dab of sticky jelly.

As soon as the thing is placed in position, the shrunken male flees and goes to recruit, after his disastrous prowess, on a slice of pear. The other, not at all troubled in spite of her heavy load, wanders about on the trellis-work of the cage, taking very short steps as she slightly raises her raspberry, this enormous burden, equal in bulk to half the creature's abdomen.

Two or three hours pass in this way. Then the Ephippiger curves herself into a ring and with her mandibles picks off particles of the nippled capsule, without bursting it, of course, or allowing the contents to flow forth. She strips its surface by removing tiny shreds, which she chews in a leisurely fashion and swallows. This fastidious consuming by atoms is continued for a whole afternoon. Next day the raspberry has disappeared; the whole of it has been gulped down during the night.

At other times the end is less quick and, above all, less repulsive. I have kept a note of an Ephippiger who was dragging her satchel along the ground and nibbling at it from time to time. The soil is uneven and rugged, having been recently turned over with the blade of a knife. The raspberry-like capsule picks up grains of sand and little clods of earth, which increase the weight of the load considerably, though the insect appears to pay no heed to it. Sometimes the carting becomes laborious, because the load sticks to some bit of earth that refuses to move. In spite of the efforts made to release the thing, it does not become detached from the point where it hangs under the ovipositor, thus proving that it possesses no small power of adhesion.

All through the evening, the Ephippiger roams about aimlessly, now on the wire-work, anon on the ground, wearing a preoccupied air. Oftener still she stands without moving. The capsule withers a little, but does not decrease notably in volume. There are no more of those mouthfuls which the Ephippiger snatched at the beginning; and the little that has already been removed affects only the surface.

Next day, things are as they were. There is nothing new, nor on the morrow either, save that the capsule withers still more, though its two red dots remain almost as bright as at first. Finally, after sticking on for forty-eight hours, the whole thing comes off without the insect's intervention.

The capsule has yielded its contents. It is a dried-up wreck, shrivelled beyond recognition, left lying in the gutter and doomed sooner or later to become the booty of the Ants. Why is it thus abandoned when, in other cases, I have seen the Ephippiger so greedy for the morsel? Perhaps because the nuptial dish had become too gritty with grains of sand, so unpleasant to the teeth.

Another Locustid, the Phaneroptera who carries a short yataghan bent into a reaping-hook (*P. falcata,* Scop.), has made up to me in part for my stud troubles. Repeatedly, but always under conditions which did not allow of completing my observation, I have caught her carrying the fertilizing-concern under the base of her saber. It is a diaphanous, oval phial, measuring three or four millimeters[4] and hanging from a crystal thread, a neck almost as long as the distended part. The insect does not touch it, but leaves the phial to dry up and shrivel where it is.[5]

Let us be content with this. These five examples, furnished by such different genera, Decticus, Analota, Grasshopper, Ephippiger and Phaneroptera, prove that the Locustid, like the Scolopendra and the Cephalopod, is a belated representative of the manners of antiquity, a valuable specimen of the genetic eccentricities of olden times.

NOTES

1. The 14th of July, the anniversary of the fall of the Bastille. – *Translator's Note.*

2. The order of insects comprising the Grasshoppers, Locusts, Crickets, Cockroaches, Mantes and Earwigs. The Cicada belongs to the order of Homoptera. – *Translator's Note.*

3. The author was obviously thinking of the Englishman's saddle of mutton and red-currant jelly. The mistake has been repeated much nearer to these shores. I have in mind the true story of an Irish king's counsel singing the praises of another, still among us, who had married an English wife and who, in the course of an extensive practice in the House of Lords, spent much of his time in England:

   "Ah, — — is a real gentleman! He speaks with an English accent, quotes Euripides in the original Latin and takes jam with his meat."

   I venture to think that Fabre, in the gentleness of his heart, would have forgiven his translator for quoting this flippant anecdote. I have no other excuse. – *Translator's Note.*

4. .117 to .156 inch. – *Translator's Note.*

5. Fuller details on this curious subject would be out of place in a book in which anatomy and physiology cannot always speak quite freely. They will be found in my essay on the Locustidæ which appeared in the *Annales des sciences naturelles,* 1896. – *Author's Note.*

33

Copris lunaris

HEREDITY

Facts which I have set forth elsewhere prove that certain Dung-beetles[1] make an exception to the rule of paternal indifference – a general rule in the insect world – and know something of domestic cooperation. The father works with almost the same zeal as the mother in providing for the settlement of the family. Whence do these favored ones derive a gift that borders on morality?

One might suggest the cost of installing the youngsters. Once they have to be furnished with a lodging and to be left the wherewithal to live, is it not an advantage, in the interests of the race, that the father should come to the mother's assistance? Work divided between the two will ensure the comfort which solitary work, its strength overtaxed, would deny. This seems excellent reasoning; but it is much more often contradicted than confirmed by the facts. Why is the Sisyphus a hard-working paterfamilias and the Sacred Beetle[2] an idle vagabond? And yet the two pill-rollers practice the same industry and the same method of rearing their young.

Why does the Lunary Copris know what his near kinsman, the Spanish Copris,[3] does not? The first assists his mate, never forsakes her. The second seeks a divorce at an early stage and leaves the nuptial roof before the children's rations are massed and kneaded into shape. Nevertheless, on both sides, there is the same big outlay on a cellarful of egg-shaped pills, whose neat rows call for long and watchful supervision. The similarity of the produce leads one to believe in similarity of manners; and this is a mistake.

Let us turn elsewhere, to the Wasps and Bees, who unquestionably come first in the laying-up of a heritage for their offspring. Whether the treasure hoarded for the benefit of the sons be a pot of honey or a bag of game, the father never takes the smallest part in the work.

He does not so much as give a sweep of the broom when it comes to tidying the outside of the dwelling. To do nothing is his invariable rule. The bringing-up of the family, therefore, however expensive it may be in certain cases, has not given rise to the instinct of paternity. Then where are we to look for a reply?

Let us make the question a wider one. Let us leave the animal, for a moment, and occupy ourselves with man. We have our own instincts, some of which take the name of genius when they attain a degree of might that towers over the plain of mediocrity. We are amazed by the unusual, springing out of flat commonplaces; we are spell-bound by the luminous speck shining in the wonted darkness. We admire; and, failing to understand whence came those glorious harvests in this one or in that, we say of them:

"They have the gift."

A goatherd amuses himself by making combinations with heaps of little pebbles. He becomes an astoundingly quick and accurate reckoner without other aid than a moment's reflection. He terrifies us with the conflict of enormous numbers which blend in an orderly fashion in his mind, but whose mere statement overwhelms us by its inextricable confusion. This marvellous arithmetical juggler has an instinct, a genius, a gift for figures.

A second, at the age when most of us delight in tops and marbles, leaves the company of his boisterous playmates and listens to the echo of celestial harps singing within him. His head is a cathedral filled with the strains of an imaginary organ. Rich cadences, a secret concert heard by him and him alone, steep him in ecstasy. All hail to that predestined one who, some day, will rouse our noblest emotions with his musical chords. He has an instinct, a genius, a gift for sounds.

A third, a child who cannot yet eat his bread and jam without smearing his face all over, takes a delight in fashioning clay into little figures that are astonishingly lifelike for all their artless awkwardness. He takes a knife and makes the briar-root grin into all sorts of entertaining masks; he carves boxwood in the semblance of a horse or sheep; he engraves the effigy of his dog on sandstone. Leave him alone; and, if Heaven second his efforts, he may become a famous sculptor. He has an instinct, a gift, a genius for form.

And so with others in every branch of human activity: art and science, industry and commerce, literature and philosophy. We have within us, from the start, that which will distinguish us from the vulgar herd. Now to what do we owe this distinctive character? To some throwback of atavism, men tell us. Heredity, direct in one case, remote in another, hands it down to us, increased or modified by time. Search the records of the family and you will discover the source of the genius, a mere trickle at first, then a stream, then a mighty river.

The darkness that lies behind that word heredity! Metaphysical science has tried to throw a little light upon it and has succeeded only in making unto itself a barbarous jargon, leaving obscurity more obscure than before. As for us, who hunger after lucidity, let us relinquish abstruse theories to whoso delights in them and confine our ambition to observable facts, without pretending to explain the quackery of the plasma. Our method certainly will not reveal to us the origin of instinct; but it will at least show us where it would be a waste of time to look for it.

In this sort of research, a subject known through and through, down to its most intimate peculiarities, is indispensable. Where shall we find that subject? There would be a host of them and magnificent ones, if it were possible to read the sealed pages of others' lives; but no one can sound an existence outside his own and even then he can think himself lucky if a retentive memory and the habit of reflection give his soundings the proper accuracy. As none of us is able to project himself into another's skin, we must needs, in considering this problem, remain inside our own.

To talk about one's self is hateful, I know. The reader must have the kindness to excuse me for the sake of the study in hand. I shall take the silent Beetle's place in the witness-box, cross-examining myself in all simplicity of soul, as I do the animal, and asking myself whence that one of my instincts which stands out above the others is derived.

Since Darwin bestowed upon me the title of "incomparable observer," the epithet has often come back to me, from this side and from that, without my yet understanding what particular merit I have shown. It seems to me so natural, so much within everybody's scope, so absorbing to interest one's self in everything that swarms around us! However, let us pass on and admit that the compliment is not unfounded.

My hesitation ceases if it is a question of admitting my curiosity in matters that concern the insect. Yes, I possess the gift, the instinct that impels me to frequent that singular world; yes, I know that I am capable of spending on those studies an amount of precious time which would be better employed in making provision, if possible, for the poverty of old age; yes, I confess that I am an enthusiastic observer of the animal. How was this characteristic propensity, at once the torment and delight of my life, developed? And, to begin with, how much does it owe to heredity?

The common people have no history: persecuted by the present, they cannot think of preserving the memory of the past. And yet what surpassingly instructive records, comforting too and pious, would be the family-papers that should tell us who our forebears were and speak to us of their patient struggles with harsh fate, their stubborn efforts to build up, atom by atom, what we are to-day. No story would come up with that for individual interest. But, by the very force of things, the home is abandoned; and, when the brood has flown, the nest is no longer recognized.

I, a humble journeyman in the toilers' hive, am therefore very poor in family-recollections. In the second degree of ancestry, my facts become suddenly obscured. I will linger over them a moment for two reasons: first, to enquire into the influence of heredity; and, secondly, to leave my children yet one more page concerning them.

I did not know my maternal grandfather. This venerable ancestor was, I have been told, a process-server in one of the poorest parishes of the Rouergue.[4] He used to engross on stamped paper in a primitive spelling. With his well-filled pencase and inkhorn, he went drawing out deeds up hill and down dale, from one insolvent wretch to another more insolvent still. Amid his atmosphere of pettifoggery, this rudimentary scholar, waging battle on life's acerbities, certainly paid no attention to the insect; at most, if he met it, he would crush it under foot. The unknown animal, suspected of evil-doing, deserved no further enquiry. Grandmother, on her side, apart from her housekeeping and her beads, knew still less about anything. She looked on the alphabet as a set of hieroglyphics only fit to spoil your sight for nothing, unless you were scribbling on paper bearing the government stamp. Who in the world, in her day, among the small folk, dreamt of knowing how to read and write? That luxury was reserved for the attorney, who himself made but a sparing use of it. The insect, I need hardly say, was the least of her cares. If sometimes, when rinsing her salad at the tap, she found a Caterpillar on the lettuce-leaves, with a start of fright she would fling the

Papilio machaon

loathsome thing away, thus cutting short relations reputed dangerous. In short, to both my maternal grandparents, the insect was a creature of no interest whatever and almost always a repulsive object, which one dared not touch with the tip of one's finger. Beyond a doubt, my taste for animals was not derived from them.

I have more precise information regarding my grandparents on the father's side,[5] for their green old age allowed me to know them both. They were people of the soil, whose quarrel with the alphabet was so great that they had never opened a book in their lives; and they kept a lean farm on the cold granite ridge of the Rouergue table-land. The house, standing alone among the heath and broom, with no neighbor for many a mile around and visited at intervals by the wolves, was to them the hub of the universe. But for a few surrounding villages, whither the calves were driven on fair-days, the rest was only very vaguely known by hearsay. In this wild solitude, the mossy ferns, with their quagmires oozing with iridescent pools, supplied the cows, the principal source of wealth, with rich, wet grass. In summer, on the short swards of the slopes, the sheep were penned day and night, protected from beasts of prey by a fence of hurdles propped up with pitchforks. When the grass was cropped close at one spot, the fold was shifted elsewhither. In the center was the shepherd's rolling hut, a straw cabin. Two watch-dogs, equipped with spiked collars, were answerable for tranquility if the thieving wolf appeared in the night from out the neighboring woods.

Padded with a perpetual layer of cow-dung, in which I sank to my knees, broken up with shimmering puddles of dark-brown liquid manure, the farm-yard also boasted a numerous population. Here the lambs skipped, the geese trumpeted, the fowls scratched the ground and the sow grunted with her swarm of little pigs hanging to her dugs.

The harshness of the climate did not give husbandry the same chances. In a propitious season, they would set fire to a stretch of moorland bristling with gorse and send the swing-plough across the ground enriched with the cinders of the blaze. This yielded a few acres of rye, oats and potatoes. The best corners were kept for hemp, which furnished the distaffs and spindles of the house with the material for linen and was looked upon as grandmother's private crop.

Grandfather, therefore, was, before all, a herdsman versed in matters of cows and sheep, but completely ignorant of aught else. How dumbfounded he would have been to learn that, in the remote future, one of his family would become enamored of those insignificant animals to which he had never vouchsafed a glance in his life! Had he guessed that that lunatic was myself, the scapegrace seated at the table by his side, what a smack I should have caught in the neck, what a wrathful look!

"The idea of wasting one's time with that nonsense!" he would have thundered.

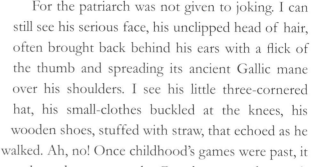

For the patriarch was not given to joking. I can still see his serious face, his unclipped head of hair, often brought back behind his ears with a flick of the thumb and spreading its ancient Gallic mane over his shoulders. I see his little three-cornered hat, his small-clothes buckled at the knees, his wooden shoes, stuffed with straw, that echoed as he walked. Ah, no! Once childhood's games were past, it would never have done to rear the Grasshopper and unearth the Dung-beetle from his natural surroundings.

Grandmother, pious soul, used to wear the eccentric head-dress of the Rouergue highlanders: a large disk of black felt, still as a plank, adorned in the middle with a crown a finger's-breadth high and hardly wider across than a six-franc piece. A black ribbon fastened under the chin maintained the equilibrium of this elegant, but unsteady circle. Pickles, hemp, chickens, curds and whey, butter; washing the clothes, minding the children, seeing to the meals of the household: say that and you have summed up the strenuous woman's round of ideas. On her left side, the distaff, with its load of flax; in her right hand, the spindle turning under a quick twist of her thumb, moistened at intervals with her tongue: so she went through life, unweariedly, attending to the order and the welfare of the house. I see her in my mind's eye particularly on winter evenings, which were more favorable to family-talk. When the hour came for meals, all of us, big and little, would take our seats round a long table, on a couple of benches, deal planks supported by four rickety legs. Each found his wooden bowl and his tin spoon in front of him. At one end of the table always stood an enormous rye-loaf, the size of a cartwheel, wrapped in a linen cloth with a pleasant smell of washing, and [there it] remained until nothing was left of it. With a vigorous stroke, grandfather would cut off enough for the needs of the moment; then he would divide the piece among us with the one knife which he alone was entitled to wield. It was now each one's business to break up his bit with his fingers and to fill his bowl as he pleased.·

Next came grandmother's turn. A capacious pot bubbled lustily and sang upon the flames in the hearth, exhaling an appetizing savor of bacon and turnips. Armed with a long metal ladle, grandmother would take from it, for each of us in turn, first the broth, wherein to soak the bread, and next the ration of turnips and bacon, partly fat and partly lean, filling the bowl to the top. At the other end of the table was the pitcher, from which the thirsty were free to drink at will. What appetites we had and what festive meals those were, especially when a cream cheese, home-made, was there to complete the banquet!

Near us blazed the huge fire-place, in which whole tree-trunks were consumed in the extreme cold weather. From a corner of that monumental, soot-glazed chimney, projected, at a convenient height, a bracket with a slate shelf, which served to light the kitchen when we sat up late. On this we burnt chips of pine-wood, selected among the most translucent, those containing the most resin. They shed over the room a lurid red light, which saved the walnut-oil in the lamp.

When the bowls were emptied and the last crumb of cheese scraped up, grandam went back to her distaff, on a stool by the chimney-corner. We children, boys and girls, squatting on our heels and putting out our hands to the cheerful fire of furze, formed a circle round her and listened to her with eager ears. She told us stories, not greatly varied, it is true, but still wonderful, for the wolf often played a part in them. I should have very much liked to see this wolf, the hero of so many tales that made our flesh creep; but the shepherd always refused to take me into his straw hut, in the middle of the fold, at night. When we had done talking about the horrid wolf, the dragon and the serpent and when the resinous splinters had given out their last gleams, we went to sleep the sweet sleep that toil gives. As the youngest of the household, I had a right to the mattress, a sack stuffed with oat-chaff. The others had to be content with straw.

I owe a great deal to you, dear grandmother: it was in your lap that I found consolation for my first sorrows. You have handed down to me, perhaps, a little of your physical vigor, a little of your love of work; but certainly you were no more accountable than grandfather for my passion for insects.

Nor was either of my own parents. My mother, who was quite illiterate, having known no teacher than the bitter experience of a harassed life, was the exact opposite of what my tastes required for their development. My peculiarity must seek its origin elsewhere: that I will swear. But I do not find it in my father, either. The excellent man, who was hard-working and sturdily-built like grandad, had been to school as a child. He knew how to write, though he took the greatest liberties with spelling; he knew how to read and understood what he read, provided the reading presented no more serious literary difficulties than occurred in the stories in the almanack. He was the first of his line to allow himself to be tempted by the town and he lived to regret it. Badly off, having but little outlet for his industry, making God knows what shifts to pick up a livelihood,[6] he went through all the disappointments of the countryman turned townsman. Persecuted by bad luck, borne down by the burden, for all his energy and good-will, he was far indeed from starting me in entomology. He had other cares, cares more direct and more serious. A good cuff or two when he saw me pinning an insect to a cork was all the encouragement that I received from him. Perhaps he was right.

The conclusion is positive: there is nothing in heredity to explain my taste for observation. You may say that I do not go far enough back. Well, what should I find beyond the grandparents where my facts come to a stop? I know, partly. I should find even more uncultured ancestors: sons

of the soil, ploughmen, sowers of rye, neat-herds; one and all, by the very force of things, of not the least account in the nice matters of observation.

And yet, in me, the observer, the enquirer into things began to take shape almost in infancy. Why should I not describe my first discoveries? They are ingenuous in the extreme, but will serve notwithstanding to tell us something of the way in which tendencies first show themselves. I was five or six years old. That the poor household might have one mouth less to feed, I had been placed in grandmother's care, as I have just been saying. Here, in solitude, my first gleams of intelligence were awakened amidst the geese, the calves and the sheep. Everything before that is impenetrable darkness. My real birth is at that moment when the dawn of personality rises, dispersing the mists of unconsciousness and leaving a lasting memory. I can see myself plainly, clad in a soiled frieze frock flapping against my bare heels; I remember the handkerchief hanging from my waist by a bit of string, a handkerchief often lost and replaced by the back of my sleeve.

There I stand one day, a pensive urchin, with my hands behind my back and my face turned to the sun. The dazzling splendor fascinates me. I am the Moth attracted by the light of the lamp. With what am I enjoying the glorious radiance: with my mouth or my eyes? That is the question put by my budding scientific curiosity. Reader, do not smile: the future observer is already practicing and experimenting. I open my mouth wide and close my eyes: the glory disappears. I open my eyes and shut my mouth: the glory reappears. I repeat the performance, with the same result. The question's solved: I have learned by deduction that I see the sun with my eyes. Oh, what a discovery! That evening, I told the whole house all about it. Grandmother smiled fondly at my simplicity: the others laughed at it. 'Tis the way of the world.

Another find. At nightfall, amidst the neighboring bushes, a sort of jingle attracted my attention, sounding very faintly and softly through the evening silence. Who is making that noise? Is it a little bird chirping in his nest? We must look into the matter and that quickly. True, there is the wolf, who comes out of the woods at this time, so they tell me. Let's go all the same, but not too far: just there, behind that clump of broom. I stand on the look-out for long, but all in vain. At the faintest sound of movement in the brushwood, the jingle ceases. I try again next day and the day after. This time, my stubborn watch succeeds. Whoosh! A grab of my hand and I hold the singer. It is not a bird; it is a kind of Grasshopper whose hind-legs my playfellows have taught me to like: a poor recompense for my prolonged ambush. The best part of the business is not the two haunches with the shrimpy flavor, but what I have just learned. I now know, from personal observation, that the Grasshopper sings. I did not publish my discovery, for fear of the same laughter that greeted my story about the sun.

Oh, what pretty flowers, in a field close to the house! They seem to smile to me with their great violet eyes. Later on, I see, in their place,

bunches of big red cherries. I taste them. They are not nice and they have no stones. What can those cherries be? At the end of the summer, grandfather comes with a spade and turns my field of observation topsy-turvy. From under ground there comes, by the basketful and sackful, a sort of round root. I know that root; it abounds in the house; time after time I have cooked it in the peat-stove. It is the potato. Its violet flower and its red fruit are pigeon-holed for good and all in my memory.

With an ever-watchful eye for animals and plants, the future observer, the little six-year-old monkey, practiced by himself, all unawares. He went to the flower, he went to the insect, even as the Large White Butterfly goes to the cabbage and the Red Admiral to the thistle. He looked and enquired, drawn by a curiosity whereof heredity did not know the secret. He bore within him the germ of a faculty unknown to his family; he kept alive a glimmer that was foreign to the ancestral hearth. What will become of that infinitesimal spark of childish fancy? It will die out, beyond a doubt, unless education intervene, giving it the fuel of example, fanning it with the breath of experience. In that case, schooling will explain what heredity leaves unexplained.

NOTES

1. The Lunary Copris and the Bison Oritis, the essay on whom has not yet appeared in English; *Minotaurus typhœus*, for whom see *The Life and Love of the Insect:* chap. x.; and the Sisyphus, for whom see *Social Life in the Insect World:* chap. xii. – *Translator's Note.*

2. See *Insect Life:* chap. i.; and *The Life and Love of the Insect:* chaps. i to iv. – *Translator's Note.*

3. See *The Life and Love of the Insect:* chap. v. – *Translator's Note.*

4. A district of the province of Guienne, having Rodez for its capital. The author's maternal grandfather, Salgues by name, was the *huissier*, or, as we should say, sheriff's officer, of Saint Léons. – *Translator's Note.*

5. Pierre Jean Fabre, son of Pierre Fabre, a peasant proprietor, and of Anne Fages, his wife, and Élisabeth Poujade, daughter of Antoine Poujade and Françoise Azémar, his wife. They were married in 1791. Pierre Fabre, a laborer, father to Pierre Jean Fabre and grandfather to Antoine Fabre, the father of Jean Henri Casimir Fabre, our author, was the son of Jean Fabre and of Françoise Desmazes, his wife, and was married in 1759 to Anne Fages, daughter of Pierre Fages and of Anne Baumelou, his wife. – *Translator's Note.*

6. The author's father kept a café in more than one small town in the south of France. – *Translator's Note.*

Melalontha melalontha
(Cockchafer)

MY SCHOOLING

I am back in the village, in my father's house. I am now seven years old; and it is high time that I went to school. Nothing could have turned out better: the master is my godfather. What shall I call the room in which I was to become acquainted with the alphabet? It would be difficult to find the exact word, because the room served for every purpose. It was at once a school, a kitchen, a bed-room, a dining-room and, at times, a chicken-house and a piggery. Palatial schools were not dreamt of in those days; any wretched hovel was thought good enough.

A broad fixed ladder led to the floor above. Under the ladder stood a big bed in a boarded recess. What was there upstairs? I never quite knew. I would see the master sometimes bring down an armful of hay for the ass, sometimes a basket of potatoes which the housewife emptied into the pot in which the little porkers' food was cooked. It must have been a loft of sorts, a storehouse of provisions for man and beast. Those two apartments composed the whole building.

To return to the lower one, the schoolroom: a window faces south, the only window in the house, a low, narrow window whose frame you can touch at the same time with your head and both your shoulders. This sunny aperture is the only lively spot in the dwelling, it overlooks the greater part of the village, which straggles along the slopes of a slanting valley. In the window-recess is the master's little table.

The opposite wall contains a niche in which stands a gleaming copper pail full of water. Here the parched children can relieve their thirst when they please, with a cup left within their reach. At the top of the niche are a few shelves bright with pewter plates, dishes and drinking-vessels, which are taken down from their sanctuary on great occasions only.

More or less everywhere, at any spot which the light touches, are crudely-colored pictures, pasted on the walls. Here is Our Lady of the Seven Dolours, the disconsolate Mother of God, opening her blue cloak to show her heart pierced with seven daggers. Between the sun and moon, which stare at you with their great, round eyes, is the Eternal Father, whose robe swells as though puffed out with the storm. To the right of the window, in the embrasure, is the Wandering Jew. He wears a three-cornered hat, a large, white leather apron, hobnailed shoes and [carries] a stout stick. "Never was such a bearded man seen before or after," says the legend that surrounds the picture. The draughtsman has not forgotten this detail: the old man's beard spreads in a snowy avalanche over the apron and comes down to his knees. On the left is Geneviève of Brabant, accompanied by the roe, with fierce Golo hiding in the bushes, sword in hand. Above hangs *The Death of Mr. Credit*, slain by defaulters at the door of his inn; and so on and so on, in every variety of subject, at all the unoccupied spots of the four walls.

I was filled with admiration of this picture-gallery, which held one's eyes with its great patches of red, blue, green and yellow. The master, however, had not set up his collection with a view to training our minds and hearts. That was the last and least of the worthy man's ambitions. An artist in his fashion, he had adorned his house according to his taste; and we benefited by the scheme of decoration.

While the gallery of halfpenny pictures made me happy all the year round, there was another entertainment which I found particularly attractive in winter, in frosty weather, when the snow lay long on the ground. Against the far wall stands the fireplace, as monumental in size as at my grandmother's. Its arched cornice occupies the whole width of the room, for the enormous redoubt fulfils more than one purpose. In the middle is the hearth, but, on the right and left, are two breast-high recesses, half wood and half stone. Each of them is a bed, with a mattress stuffed with chaff of winnowed corn. Two sliding planks serve as shutters and close the chest if the sleeper would be alone. This dormitory, sheltered under the chimney-mantel, supplies couches for the favored ones of the house, the two boarders. They must lie snug in there at night, with their shutters closed, when the north-wind howls at the mouth of the dark valley and sends the snow awhirl. The rest is occupied by the hearth and its accessories: the three-legged stools; the salt-box, hanging against the wall to keep its contents dry; the heavy shovel which it takes two hands to wield; lastly, the bellows similar to those with which I used to blow out my cheeks in grandfather's house. They consist of a mighty branch of pine,

hollowed throughout its length with a red-hot iron. By means of this channel, one's breath is applied, from a convenient distance, to the spot which is to be revived. With a couple of stones for supports, the master's bundle of sticks and our own logs blaze and flicker, each of us having to bring a log of wood in the morning, if he would share in the treat.

For that matter, the fire was not exactly lit for us, but, above all, to warm a row of three pots in which simmered the pigs' food, a mixture of potatoes and bran. That, despite the tribute of a log, was the real object of the brushwood-fire. The two boarders, on their stools, in the best places, and we others, sitting on our heels, formed a semicircle around those big cauldrons, full to the brim and giving off little jets of steam, with puff-puff-puffing sounds. The bolder among us, when the master's eyes were engaged elsewhere, would dig a knife into a well-cooked potato and add it to their bit of bread; for I must say that, if we did little work in my school, at least we did a deal of eating. It was the regular custom to crack a few nuts and nibble at a crust while writing our page or setting out our rows of figures.

We, the smaller ones, in addition to the comfort of studying with our mouths full, had every now and then two other delights, which were quite as good as cracking nuts. The back-door communicated with the yard where the hen, surrounded by her brood of chicks, scratched at the dung-hill, while the little porkers, of whom there were a dozen, wallowed in their stone trough. This door would open sometimes to let one of us out, a privilege which we abused, for the sly ones among us were careful not to close it on returning. Forthwith, the porkers would come running in, one after the other, attracted by the smell of the boiled potatoes. My bench, the one where the youngsters sat, stood against the wall, under the copper pail to which we used to go for water when the nuts had made us thirsty, and was right in the way of the pigs. Up they came trotting and grunting, curling their little tails; they rubbed against our legs; they poked their cold pink snouts into our hands in search of a scrap of crust; they questioned us with their sharp little eyes to learn if we happened to have a dry chestnut for them in our pockets. When they had gone the round, some this way and some that, they went back to the farmyard, driven away by a friendly flick of the master's handkerchief. Next came the visit of the hen, bringing her velvet-coated chicks to see us. All of us eagerly crumbled a little bread for our pretty visitors. We vied with one another in calling them to us and tickling with our fingers their soft and downy backs. No, there was certainly no lack of distractions.

What could we learn in such a school as that! Let us first speak of the young ones, of whom I was one. Each of us had, or rather was supposed to have, in his hands a little penny book, the alphabet, printed on grey paper. It began, on the cover, with a pigeon, or something like it. Next came a cross, followed by the letters in their order. When we turned over, our eyes encountered the terrible *ba, be, bi, bo, bu,* the stumbling-block of most of us. When we had mastered that formidable page, we were considered to know how to read and were admitted among the big ones. But, if the little book was to be of any use, the least that was required was that the master should interest himself in us to some extent and show us how to set about things. For this, the worthy man, too much taken up with the big ones, had not the time. The famous alphabet with the pigeon was thrust upon us only to give us the air of scholars. We were to contemplate it on our bench, to decipher it with the help of our next neighbor, in case he might know one or two of the letters. Our contemplation came to nothing, being every moment disturbed by a visit to the potatoes in the stewpots, a quarrel among playmates about a marble, the grunting invasion of the porkers or the arrival of the chicks. With the aid of these distractions, we would wait patiently until it was time for us to go home. That was our most serious work.

The big ones used to write. They had the benefit of the small amount of light in the room, by the narrow window where the Wandering Jew and ruthless Golo faced each other, and of the large and only table with its circle of seats. The school supplied nothing, not even a drop of ink; every one had to come with a full set of utensils. The inkhorn of those days, a relic of the ancient pencase of which Rabelais speaks, was a long cardboard box divided into two stages. The upper compartment held the pens, made of goose- or turkey-quills trimmed with a penknife; the lower contained, in a tiny well, ink made of soot mixed with vinegar.

The master's great business was to mend the pens – a delicate work, not without danger for inexperienced fingers – and then to trace at the head of the white page a line of strokes, single letters or words, according to the scholar's capabilities. When that is over, keep an eye on the work of art which is coming to adorn the copy! With what undulating movements of the wrist does the hand, resting on the little finger, prepare and plan its flight! All at once, the hand starts off, flies, whirls; and, lo and behold, under the line of writing is unfurled a garland of circles, spirals and flourishes, framing a bird with outspread wings, the whole, if you please, in red ink, the only kind worthy of such a pen. Large and small, we stood awestruck in the presence of these marvels. The family, in the evening, after supper, would pass from hand to hand the masterpiece brought back from school:

"What a man!" was the comment. "What a man, to draw you a Holy Ghost with a stroke of the pen!"

What was read at my school? At most, in French, a few selections from sacred history. Latin recurred oftener, to teach us to sing vespers properly. The more advanced pupils tried to decipher manuscript, a deed of sale, the hieroglyphics of some scrivener.

And history, geography? No one ever heard of them. What difference did it make to us whether the earth was round or square! In either case, it was just as hard to make it bring forth anything.

And grammar? The master troubled his head very little about that; and we still less. We should have been greatly surprised by the novelty and the forbidding look of such words in the grammatical jargon as substantive, indicative and subjunctive. Accuracy of language, whether of speech or writing, must be learned by practice. And none of us was troubled by scruples in this respect. What was the use of all these subtleties, when, on coming out of school, a lad simply went back to his flock of sheep!

And arithmetic? Yes, we did a little of this, but not under that learned name. We called it sums. To put down rows of figures, not too long, add them and subtract them one from the other was more or less familiar work. On Saturday evenings, to finish up the week, there was a general orgy of sums. The top boy stood up and, in a loud voice, recited the multiplication-table up to twelve times. I say twelve times, for in those days, because of our old duodecimal measures, it was the custom to count as far as the twelve-times table, instead of the ten times of the metric system. When this recital was over, the whole class, the little ones included, took it up in chorus, creating such an uproar that chicks and porkers took to flight if they happened to be there. And this went on to twelve times twelve, the first in the row starting the next table and the whole class repeating it as loud as it could yell. Of all that we were taught in school, the multiplication-table was what we knew best, for this noisy method ended by dinning the different numbers into our ears. This does not mean that we became skilful reckoners. The cleverest of us easily got muddled with the figures to be carried in a multiplication-sum. As for division, rare indeed were they who reached such heights. In short, the moment a problem, however insignificant, had to be solved, we had recourse to mental gymnastics much rather than to the learned aid of arithmetic.

When all is said, our master was an excellent man who could have kept school very well but for his lack of one thing; and that was time. He devoted to us all the little leisure which his numerous functions left him. And, first of all, he managed the property of an absentee landowner, who only occasionally set foot in the village. He had under his care an old castle with four towers, which had become so many pigeon-houses; he directed the getting-in of the hay, the walnuts, the apples and the oats. We used to help him during the summer, when the school, which was well-attended in winter, was almost deserted. All that remained, because they were not yet big enough to work in the fields, were a few children, including him who was one day to set down these memorable facts. Lessons at that time were less dull. They were often given on the hay or on the straw; oftener still, lesson-time was spent in cleaning out the dove-cote or stamping on the snails that had sallied in rainy weather from their fortresses, the tall box borders of the garden belonging to the castle.

Our master was a barber. With his light hand, which was so clever at beautifying our copies with curlicue birds, he shaved the notabilities of the place: the mayor, the parish-priest, the notary. Our master was a bell-ringer. A wedding or a christening interrupted the lessons: he had to ring a peal. A gathering storm gave us a holiday: the great bell must be tolled to ward off the lightning and the hail. Our master was a choir-singer. With his mighty voice, he filled the church when he led the *Magnificat* at vespers. Our master wound up and regulated the village-clock. This was his proudest function. Giving a glance at the sun, to ascertain the time more or less nearly, he would climb to the top of the steeple, open a huge cage of rafters and find himself in a maze of wheels and springs whereof the secret was known to him alone.

With such a school and such a master and such examples, what will become of my embryo tastes, as yet so imperceptible? In that environment, they seem bound to perish, stifled for ever. Yet no, the germ has life; it works in my veins, never to leave them again. It finds nourishment everywhere, down to the cover of my penny alphabet, embellished with a crude picture of a pigeon which I study and contemplate much more zealously than the ABC. Its round eye, with its circlet of dots, seems to smile upon me. Its wing, of which I count the feathers one by one, tells me of flights on high, among the beautiful clouds; it carries me to the beeches raising their smooth trunks above a mossy carpet studded with white mushrooms that look like eggs dropped by some vagrant hen; it takes me to the snow-clad peaks where the birds leave the starry print of their red feet. He is a fine fellow, my pigeon-friend: he consoles me for the woes hidden behind the cover of my book. Thanks to him, I sit quietly on my bench and wait more or less till school is over.

School out of doors has other charms. When the master takes us to kill the snails in the box borders, I do not always scrupulously fulfil my office as an exterminator. My heel sometimes hesitates before coming down upon the handful which I have gathered. They are so pretty! Just think, there are yellow ones and pink, white ones and brown, all with dark spiral streaks. I fill my pockets with the handsomest, so as to feast my eyes on them at my leisure.

On hay-making days in the master's field, I strike up an acquaintance with the Frog. Flayed and stuck at the end of a split stick, he serves as live bait to tempt the crayfish to come out of his retreat by the brook-side. On the alder-trees I catch the Hoplia, the splendid Scarab who pales the azure of the heavens. I pick the narcissus and learn to gather, with the tip of my tongue, the tiny drop of honey that lies right at the bottom of the cleft corolla. I also learn that too-long indulgence in this feast brings a headache; but this discomfort in no way impairs my admiration for the glorious white flower, which wears a narrow red collar at the throat of its funnel.

When we go to beat the walnut-trees, the barren grass-plots provide me with Locusts spreading their wings, some into a blue fan, others into a red. And thus the rustic school, even in the heart of winter, furnished continuous food for my interest in things. There was no need for precept and example: my passion for animals and plants made progress of itself.

What did not make progress was my acquaintance with my letters, greatly neglected in favor of the pigeon. I was still at the same stage, hopelessly behindhand with the untractable alphabet, when my father, by a chance inspiration, brought me home from the town what was destined to give me a start along the road of reading. Despite the not insignificant part which it played in my intellectual awakening, the purchase was by no means a ruinous one. It was a large print, price six farthings, colored and divided into compartments in which animals of all sorts taught the A B C by means of the first letters of their names.

Where should I keep the precious picture? As it happened, in the room set apart for the children at home, there was a little window like the one in the school, opening in the same way out of a sort of recess and in the same way overlooking most of the village. One was on the right, the other on the left of the castle with the pigeon-house towers; both afforded an equally good view of the heights of the slanting valley. I was able to enjoy the school-window only at rare intervals, when the master left his little table; the other was at my disposal as often as I liked. I spent long hours there, sitting on a little fixed window-seat.

The view was magnificent. I could see the ends of the earth, that is to say, the hills that blocked the horizon, all but a misty gap through which the brook with the crayfish flowed under the alders and willows. High up on the sky-line, a few wind-battered oaks bristled on the ridges; and beyond there lay nothing but the unknown, laden with mystery.

At the back of the hollow stood the church, with its three steeples and its clock; and, a little higher, the village-square, where a spring, fashioned into a fountain, gurgled from one basin into another, under a wide arched roof. I could hear from my window the chatter of the women washing their clothes, the strokes of their beaters, the rasping of the pots scoured with sand and vinegar. Sprinkled over the slopes are little houses with their garden-patches in terraces banked up by tottering walls, which bulge under the thrust of the earth. Here and there are very steep lanes, with the dents of the rock forming a natural pavement. The mule, sure-footed though he be, would hesitate to enter these dangerous passes with his load of branches.

Farther on, beyond the village, half-way up the hills, stood the great ever-so-old lime-tree, the *Tel*, as we used to call it, whose sides, hollowed out by the ages, were the favorite hiding-places of us children at play. On fair-days, its immense, spreading foliage cast a wide shadow over the herds of oxen and sheep. Those solemn days, which only came once a year, brought me a few ideas from without: I learned that the world did not end with my amphitheater of hills. I saw the inn-keeper's wine arrive on

A Ane

B Boeuf

C Canard

G Grenouille

H Herisson

I Insecte

M Morse

N Nautilus

O Ours

T Tigre

U Urson

V Vanneau

D Dinde

E Eléphant

F Fouine

J Jaguar

K Kudu

L Loup

P Pingouin

R Raton laveur

S Sanglier

W Wapiti

Y Yak

Z Zebu

muleback and in goat-skin bottles. I hung about the market-place and watched the opening of jars full of stewed pears, the setting-out of baskets of grapes, an almost unknown fruit, the object of eager covetousness. I stood and gazed in admiration at the roulette-board on which, for a sou, according to the spot at which its needle stopped on a circular row of nails, you won a pink poodle made of barley-sugar, or a round jar of aniseed sweets, or, much oftener, nothing at all. On a piece of canvas on the ground, rolls of printed calico with red flowers were displayed to tempt the girls. Close by rose a pile of beech-wood clogs, tops and box-wood flutes. Here the shepherds chose their instruments, trying them by blowing a note or two. How new it all was to me! What a lot of things there were to see in this world! Alas, that wonderful time was of but short duration! At night, after a little brawling at the inn, it was all over; and the village returned to silence for a year.

But I must not linger over these memories of the dawn of life. We were speaking of the memorable picture brought from town. Where shall I keep it, to make the best use of it? Why, of course, it must be pasted on the embrasure of my window! The recess, with its seat, shall be my study-cell; here I can feast my eyes by turns on the big lime-tree and the animals of my alphabet. And this was what I did.

And now, my precious picture, it is our turn, yours and mine. You began with the sacred beast, the ass, whose name, with a big initial, taught me the letter A. The *bœuf*, the ox, stood for B; the *canard*, the duck, told me about C; the *dindon*, the turkey, gave me the letter D. And so on with the rest. A few compartments, it is true, were lacking in clearness. I had no friendly feeling for the hippopotamus, the kamichi, or horned screamer, and the zebu, who aimed at making me say H, K, and Z. Those outlandish beasts, which failed to give the abstract letter the support of a recognized reality, caused me to hesitate for a time over their recalcitrant consonants. No matter: father came to my aid in difficult cases; and I made such rapid progress that, in a few days, I was able to turn in good earnest the pages of my little pigeon-book, hitherto so undecipherable. I was initiated; I knew how to spell. My parents marvelled. I can explain this unexpected progress to-day. Those speaking pictures, which brought me amongst my friends the beasts, were in harmony with my instincts. If the animal has not fulfilled all that it promised in so far as I am concerned, I have at least to thank it for teaching me to read. I should have succeeded by other means, I do not doubt, but not so quickly nor so pleasantly. Animals for ever!

Luck favored me a second time. As a reward for my prowess, I was given La Fontaine's Fables, in a popular, cheap edition, crammed with pictures, small, I admit, and very inaccurate, but still delightful. Here were the crow, the fox, the wolf, the magpie, the frog, the rabbit, the ass, the dog, the cat: all persons of my acquaintance. The glorious book was immensely to my taste, with its skimpy illustrations in which the animal walked and talked. As to understanding what it said, that was another story! Never mind, my lad! Put together syllables that say nothing to you as yet; they will speak to you later and La Fontaine will always remain your friend.

I come to the time when I was ten years old and at Rodez College. My functions as a serving-boy in the chapel entitled me to free instruction as a day-boarder. There were four of us in white surplices and red skull-caps and cassocks. I was the youngest of the party and did little more than walk on. I counted as a unit; and that was about all, for I was never certain when to ring the bell or move the missal. I was all of a tremble when we gathered two on this side and two on that, with genuflexions, in the middle of the sanctuary, to intone the *Domine, salvum fac regem* at the end of mass. Let me make a confession: tongue-tied with shyness, I used to leave it to the others.

Nevertheless, I was well thought of, for, in the school, I cut a good figure in composition and translation. In that classical atmosphere, there was talk of Procas, King of Alba, and of his two sons, Numitor and Amulius. We heard of Cynœgirus, the strong-jawed man, who, having lost his two hands in battle, seized and held a Persian galley with his teeth, and of Cadmus the Phœnician, who sowed a dragon's teeth as though they were beans and gathered his harvest in the shape of a host of armed men, who killed one another as they rose up from the ground. The only one who survived the slaughter was one as tough as leather, presumably the son of the big back grinder.

Had they talked to me about the man in the moon, I could not have been more startled. I made up for it with my animals, which I was far from forgetting amid this phantasmagoria of heroes and demigods. While honoring the exploits of Cadmus and Cynœgirus, I hardly ever failed, on Sundays and Thursdays,[1] to go and see if the cowslip or the yellow daffodil was making its appearance in the meadows, if the Linnet was hatching on the juniper-bushes, if the Cockchafers were plopping down from the wind-shaken poplars. Thus was the sacred spark kept aglow, ever brighter than before.

By easy stages, I came to Virgil and was very much smitten with Melibœus, Corydon, Menalcas, Damœtas and the rest of them. The scandals of the ancient shepherds fortunately passed unnoticed; and within the frame in which the characters moved were exquisite details concerning the Bee, the Cicada, the Turtle-dove, the Crow, the Nanny-goat and the golden broom. A veritable delight were these stories of the fields, sung in sonorous verse; and the Latin poet left a lasting impression on my classical recollections.

Then, suddenly, good-bye to my studies, good-bye to Tityrus and Menalcas! Ill-luck is swooping down on us, relentlessly. Hunger threatens us at home. And now, boy, put your trust in God; run about and earn your penn'orth of potatoes as best you can. Life is about to become a hideous inferno. Let us pass quickly over this phase.

Amid this lamentable chaos, my love for the insect ought to have gone under. Not at all. It would have survived the raft of the *Medusa*. I still remember a certain Pine Cockchafer met for the first time. The plumes on her antennæ, her pretty pattern of white spots on a dark-brown ground were as a ray of sunshine in the gloomy wretchedness of the day.

To cut a long story short: good fortune, which never abandons the brave, brought me to the primary normal school at Vaucluse, where I was assured food: dried chestnuts and chick-peas. The principal, a man of broad views, soon came to trust his new assistant. He left me practically a free hand, so long as I satisfied the school curriculum, which was very modest in those days. Possessing a smattering of Latin and grammar, I was a little ahead of my fellow-pupils. I took advantage of this to get some order into my vague knowledge of plants and animals. While a dictation-lesson was being corrected around me, with generous assistance from the dictionary, I would examine, in the recesses of my desk, the oleander's fruit, the snap-dragon's seed-vessel, the Wasp's sting and the Ground-beetle's wing-case.

With this foretaste of natural science, picked up haphazard and by stealth, I left school more deeply in love than ever with insects and flowers. And yet I had to give it all up. That wider education, which would have to be my source of livelihood in the future, demanded this imperiously. What was I to take in hand to raise me above the primary school, whose staff could barely earn their bread in those days? Natural history could not bring me anywhere. The educational system of the time kept it at a distance, as unworthy of association with Latin and Greek. Mathematics remained, with its very simple equipment: a blackboard, a bit of chalk and a few books.

So I flung myself with might and main into conic sections and the calculus: a hard battle, if ever there was one, without guides or counsellors, face to face for days on end with the abstruse problem which my stubborn

thinking at last stripped of its mysteries. Next came the physical sciences, studied in the same manner, with an impossible laboratory, the work of my own hands.

The reader can imagine the fate of my favorite branch of science in this fierce struggle. At the faintest sign of revolt, I lectured myself severely, lest I should let myself be seduced by some new grass, some unknown Beetle. I did violence to my feelings. My natural-history books were sentenced to oblivion, relegated to the bottom of a trunk.

And so, in the end, I am sent to teach physics and chemistry at Ajaccio College. This time, the temptation is too much for me. The sea, with its wonders, the beach, whereon the tide casts such beautiful shells, the *maquis* of myrtles, arbutus and mastic-trees: all this paradise of gorgeous nature has too much on its side in the struggle with the sine and the cosine. I succumb. My leisure-time is divided into two parts. One, the larger, is allotted to mathematics, the foundation of my academical future, as planned by myself; the other is spent, with much misgiving, in botanizing and looking for the treasures of the sea. What a country and what magnificent studies to be made, if, unobsessed by x and y, I had devoted myself wholeheartedly to my inclinations!

We are a wisp of straw, the plaything of the winds. We think that we are making for a goal deliberately chosen; destiny drives us towards another. Mathematics, the exaggerated preoccupation of my youth, did me hardly any service; and animals, which I avoided as much as ever I could, are the consolation of my old age. Nevertheless, I bear no grudge against the sine and the cosine, which I continue to hold in high esteem. They cost me many a pallid hour at one time, but they always afforded me some first-rate entertainment: they still do so, when my head lies tossing sleeplessly on its pillow.

Meanwhile, Ajaccio received the visit of a famous Avignon botanist, Requien[2] by name, who, with a box crammed with paper under his arm, had long been botanizing all over Corsica, pressing and drying specimens and distributing them to his friends. We soon became acquainted. I accompanied him in my free time on his explorations and never did the master have a more attentive disciple. To tell the truth, Requien was not a man of learning so much as an enthusiastic collector. Very few would have felt capable of competing with him when it came to giving the name or the geographical distribution of a plant. A blade of grass, a pad of moss, a scab of lichen, a thread of seaweed: he knew them all. The scientific name flashed across his mind at once. What an unerring memory, what a genius for classification amid the enormous mass of things observed! I stood aghast at it. I owe much to Requien in the domain of botany. Had death spared him longer, I should doubtless have owed more to him, for his was a generous heart, ever open to the troubles of novices.

In the following year, I met Moquin-Tandon,[3] with whom, thanks to Requien, I had already exchanged a few letters on botany. The illustrious Toulouse professor came to study on the spot the flora which he proposed to describe systematically. When he arrived, all the hotel bedrooms were reserved for the members of the general council which had been summoned; and I offered him board and lodging: a shake-down in a room overlooking the sea; fare consisting of lampreys, turbot and sea-urchins: common enough dishes in that land of Cockayne, but possessing no small attraction for the naturalist, because of their novelty. My cordial proposal tempted him; he yielded to my blandishments; and there we were for a fortnight chatting at table *de omni re scibili* after the botanical excursion was over.

With Moquin-Tandon, new vistas opened before me. Here it was no longer the case of a nomenclator with an infallible memory: he was a naturalist with far-reaching ideas, a philosopher who soared above petty details to comprehensive views of life, a writer, a poet who knew how to clothe the naked truth in the magic mantle of the glowing word. Never again shall I sit at an intellectual feast like that:

"Leave your mathematics," he said. "No one will take the least interest in your formulæ. Get to the beast, the plant; and, if, as I believe, the fever burns in your veins, you will find men to listen to you."

We made an expedition to the center of the island, to Monte Renoso,[4] with which I was extremely familiar. I made the scientist pick the hoary everlasting (*Helichrysum frigidum*), which makes a wonderful patch of silver; the many-headed thrift, or mouflon-grass (*Armeria multiceps*), which the Corsicans call *erba muorone*; the downy marguerite (*Leucanthemum tomosum*), which, clad in wadding, shivers amid the snows; and many other rarities dear to the botanist. Moquin-Tandon was jubilant. I, on my side, was much more attracted and overcome by his words and his enthusiasm than by the hoary everlasting. When we came down from the cold mountain-top, my mind was made up: mathematics would be abandoned.

On the day before his departure, he said to me:

"You interest yourself in shells. That is something, but it is not enough. You must look into the animal itself. I will show you how it's done."

And, taking a sharp pair of scissors from the family work-basket and a couple of needles stuck into a bit of vine-shoot which served as a makeshift handle, he showed me the anatomy of a snail in a soup-plate filled with water. Gradually he explained and sketched the organs which he spread before my eyes. This was the only, never-to-be-forgotten lesson in natural history that I ever received in my life.

It is time to conclude. I was cross-examining myself, being unable to cross-examine the silent Beetle. As far as it is possible to read within myself, I answer as follows:

"From early childhood, from the moment of my first mental awakening, I have felt drawn towards the things of nature, or, to return to our catchword, I have the gift, the bump of observation."

After the details which I have already given about my ancestors, it would be ridiculous to look to heredity for an explanation of the fact. Nor would any one venture to suggest the words or example of my masters. Of scientific education, the fruit of college-training, I had none whatever. I never set foot in a lecture-hall except to undergo the ordeal of examinations. Without masters, without guides, often without books, in spite of poverty, that terrible extinguisher, I went ahead, persisted, facing my difficulties, until the indomitable bump ended by shedding its scanty contents. Yes, they were very scanty, yet possibly of some value, if circumstances had come to their assistance. I was a born animalist. Why and how? No reply.

We thus have, all of us, in different directions and in a greater or lesser degree, characteristics that brand us with a special mark, characteristics of unfathomable origin. They exist because they exist; and that is all that any one can say. The gift is not handed down: the man of talent has a fool for a son. Nor is it acquired; but it is improved by practice. He who has not the germ of it in his veins will never possess it, in spite of all the pains of a hot-house education.

That to which we give the name of instinct when speaking of animals is something similar to genius. It is, in both cases, a peak that rises above the ordinary level. But instinct is handed down, unchanged and undiminished, throughout the sequence of a species; it is permanent and general and in this it differs greatly from genius, which is not transmissible and changes in different cases. Instinct is the inviolable heritage of the family and falls to one and all, without distinction. Here the difference ends. Independent of similarity of structure, it breaks out like genius, here or elsewhere, for no perceptible reason. Nothing causes it to be foreseen, nothing in the organization explains it. If cross-examined on this point, the Dung-beetles and the rest, each with his own peculiar talent, would answer, were we able to understand them:

"Instinct is the animal's genius."

Dung beetles

NOTES

1. The weekly half holiday in French schools. — *Translator's Note.*

2. Esprit Requien (1788–1851), a French naturalist and collector, director of the museum and botanical gardens at Avignon and author of several works on botany and conchology. — *Translator's Note.*

3. Horace Bénédict Alfred Moquin-Tandon (1804–1863), a distinguished naturalist, for twenty years director of the botanical gardens at Toulouse. He was commissioned by the French government in 1850 to compile a flora of Corsica and is the author of several important works on botany and zoology. — *Translator's Note.*

4. A mountain 7,730 feet high, about twenty-five miles from Ajaccio. — *Translator's Note.*

Pentatoma rufipes

Wheatear   *Nightingale*   *Yellowhammer*   *Butcherbird*

THE PENTATOMÆ AND
THEIR EGGS

Of the forms which life is able to bestow on her creations, that of the bird's egg is one of the simplest and loveliest. Nowhere do we find the beauty of the circle and the ellipse, the geometrical bases of organic bodies, combined with greater precision. At one of the poles is the sphere, the perfect form, capable of enclosing the greatest volume in the smallest envelope; at the other is the point of the ellipsoid, which tempers the monotonous austerities of the big end.

The color-scheme, likewise very simple, adds its graces to those of form. Some eggs display the dull white of chalk, others the translucid white of polished ivory. The Wheat-ear's are a delicate blue, like that of a sky freshly washed by a rain-storm; the Nightingale's are a dark green, like that of a pickled olive; the eggs of certain Warblers are tinted with an exquisite carnation, like that of roses still in the bud.

The Yellow-hammer scrawls an indecipherable scribble on her eggs; that is to say, the shells display mottled markings, an artistic mixture of lines and blots. The Butcher-birds encircle the large end with a speckled

Prothonotary Warbler   *Thickbilled Warbler*   *Yellow Warbler*   *Booted Warbler*

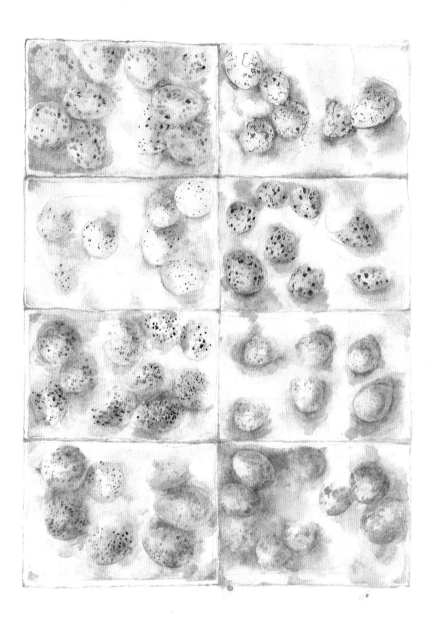

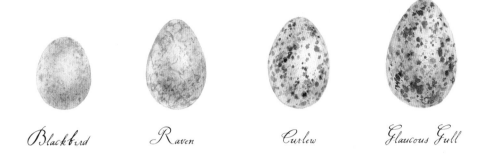

crown; the Blackbird and the Raven sprinkle brown splashes, innocent of design, on a greenish-blue ground; the Curlew and the Gull employ large spots like those on the Leopard's coat; and so with the rest; each has its specialty, its trade-mark, always designed in sober colors, the mere matching of which constitutes a merit.

With the exquisite simplicity of its geometry and its ornament, the bird's egg enchants the least cultivated eye. In return for the little services which they render me, I sometimes admit to my study certain small boys of the neighborhood, zealous searchers all. Now what do these simple-minded youngsters see in my work-room, of which they have heard all sorts of wonders? They see big, glass-fronted cupboards in which a thousand curious things are arranged, the cumbersome accumulations that gather about any one who investigates stones, plants and animals. Shells predominate.

Huddling together in mutual encouragement, my shy visitors admire the magnificent Sea-snails of every shape and color; they point a finger at this or that shell which, by the luster of its mother-of-pearl, its size and its strange protuberances, is especially conspicuous in the midst of all the rest. They gaze at my treasures and I watch their faces. I read on them surprise, amazement and nothing more.

These things out of the sea, too complex in formation to impress a novice, are mysterious objects that speak no known language. My little giddy-pates are bewildered by these corkscrew stair-cases, these scrolls and spirals and conchs, whose geometry is beyond their comprehension.

Nightingale

They are left almost cold before this display of oceanic wealth. If I could get at what lies at the back of their minds, these children would say:

"How funny!"

They would never say:

"How pretty!"

It is quite another story with the boxes in which the birds'-eggs of the district are arranged, clutch by clutch, lying on cotton wool, protected from the light. Now their cheeks flush with excitement and they whisper, in one another's ears, which they would choose of the finest group in the box. There is no amazement now, but ingenuous admiration. It is true that the egg recalls the nest and the young birds, those incomparable joys of childhood. Nevertheless, a rush of reverent emotion evoked by the beautiful may be read on their faces. The gems of the sea astound my little visitors; the simple beauty of the eggs arouses a more human ecstasy.

In the very great majority of cases, the insect's egg is far from attaining this consummate perfection, which impresses even the unaccustomed gaze. The usual shapes are the sphere, the spindle or cone, and the cylinder, with rounded ends, none of which is especially graceful, owing to the absence of harmonious combinations of curves. Many of them are dingy in color; some, by their excessive richness, form a violent contrast with the shortcomings of the germ inside. The eggs of certain Moths and Butterflies are beads of bronze or nickel. In these life seems to germinate within the rigid walls of a metal box.

If we employ the magnifying-glass, we find that ornamentation of detail is not unusual, but it is always complicated, without that nobler simplicity which constitutes true beauty. The Clythræ[1] enclose their eggs in a shell whose substance is laminated in scales like those of a hop-cone, or twisted into intersecting diagonal fillets; certain Locusts engrave their spindles, scooping out spiral

Yellowhammer

rows of little pits like those of a thimble. There is, to be sure, no lack of prettiness in all this, but how far removed is such exuberance from the noble austerity of beauty!

The insect has ovarian æsthetics of its own, which have no relation to those of the bird. I know of one case, however, in which comparison is possible. An insect of indifferent repute, a woodland Bug, the Pentatoma of the naturalists, may offer its egg for comparison with the bird's. This flat-bodied insect, emitting a horrible smell, lays masterpieces of elegant simplicity, and, at the same time, of mechanical ingenuity; it disgusts us by its cosmetic, its hair-oil; but it interests us by its egg, which is worthy to rank beside that of the bird.

I have just made a discovery on a sprig of asparagus. It is a cluster of eggs, about thirty in number, arranged in rows, in close contact, like the beads on a piece of embroidery. I recognize the eggs of a woodland Bug. The hatching took place some little time ago, for the family has not yet dispersed. The empty eggshells have remained in place without any loss of shape, except that their lids are open.

What a delightful collection of miniature vases in translucent alabaster, barely clouded with light grey! One would like to read a fairy-tale of the world of tiny things in which the fairies take tea out of such cups as these. The body of the vessel, a graceful oval cut square at the top, shows a delicate brown network of polygonal meshes. Imagine the top of a bird's egg neatly removed, making a dainty little goblet of the remainder, and you have something very like the egg of the Bug. In either case there are the same gentle curves.

Here the resemblance ceases. It is in the upper part of the egg that the insect displays its originality; its creation is a box with a lid. This slightly convex cover is ornamented, like the body of the jar, with a network of fine mesh; it is further embellished along the edge with an opal border. At the hatching it swings open as on a hinge and comes away all of a piece.

Sometimes it falls off and leaves the jar wide open; sometimes it falls back into its normal position, once more closing the jar, which looks as though it were still intact. Lastly, the mouth is surrounded by very fine, thread-like attachments. These are, as it were, rivets to hold the lid in position, so as to close the vase hermetically.

We must not overlook one exceedingly characteristic detail. Quite close to the rim, inside the shell, there is always visible, after the hatching, a mark like a broad arrow, or a capital T, with the arms deflected like those of an anchor. What is the meaning of this infinitesimal detail? Is it a latch, a sort of lock with a bolt and hasp? Is it a potter's mark, conferring a certificate of origin on the masterpiece? What a strange effort of ceramic art merely to hold the egg of a Bug!

The young ones have not yet left the battery of jars from which they recently emerged. Gathered together in a heap, they are waiting for the bath of air and sunlight to harden them before dispersing and implanting their suckers where they please. They are plump, thickset, black, with the under surface of the belly red and the sides laced with the same color. How did they get out of their jars? By what artifice did they raise the firmly-sealed lid? Let us try to find the answer to this interesting question.

It is the end of April. In the enclosure, just outside my door, the camphor-scented rosemaries are in full flower, bringing me visits from a multitude of insects which I can consult at any time. Various species of Pentatomæ abound, but do not lend themselves to precise observation, by reason of their wandering life. If I want to know exactly which egg belongs to which species or, above all, if I want to learn how the hatching is accomplished, it will not be enough to rely upon chance inspections of the flowering shrubs. It will be better to resort to rearing the insects under a wire-gauze cover.

My captives, isolated according to species and represented each by a certain number of couples, give me hardly any trouble. All they need is a cheerful sun and a bunch of rosemary daily renewed. I add to the furnishing of the cage a few leafy twigs from various bushes. The insect will choose whichever suits her as the spot for laying her eggs.

By the first fortnight in May the imprisoned Bugs have provided me with eggs in excess of my hopes, eggs at once collected, together with their support, species by species, and placed in small glass tubes, where unless I fail in vigilance, I shall easily be able to follow the delicate hatching-process.

It is really a beautiful, a most delightful collection, and would be quite worthy to figure beside the eggs of the bird, if larger dimensions came to the assistance of our feeble sight. From the moment we have to resort to the microscope, we allow the splendid to escape us. Let us magnify the Bug's egg under the lens and it will amaze us as surely as the Stonechat's sky-blue egg, and perhaps even more. What a pity that such beauty escapes our admiration by its minuteness!

The shape is never a complete ovoid: that is the bird's perquisite. The upper end of the Pentatoma's egg is always finished off with a sudden

truncation, into which a slightly convex lid is fitted, and we have before us a tiny ciborium, a delicious casket, an antique urn, a cylindrical cask with rounded ends, a full-bodied vase of Oriental porcelain, with ornaments consisting of bands, rosettes or traceries, varying according to the mother's individual taste. Always, moreover, when the egg is empty, we find a most delicate fringe of herring-boned threads running round the mouth. These are the rivets to fasten the lid, which are pushed up and back at the moment when the new-born insect is released.

Lastly, in all these egg-shells, after the hatching, we find inside them, quite close to the rim, that black mark in the shape of a broad arrow, of which we have already asked ourselves whether it is a trade-mark or a sort of lock or bolt. The future will show us how far our guesses fall short of the reality.

The eggs are never sown at random. The whole batch is laid in a close-packed group, in regular ranks of varying lengths, so that they make a sort of mosaic of beads firmly fixed to their common support, usually a leaf. They adhere so firmly that we may brush the leaf with a camel-hair pencil, or even touch them with the finger, without in any way disturbing their beautiful arrangement. After the young have gone we find the open shells still in position, like so many little jam-pots standing in rows on a market-woman's barrow.

Let me end by giving a few specific details. The eggs of the Black-horned Pentatoma (*P. nigricorne*) are cylindroid in form, the base being a segment of a sphere. The lid, bearing a broad white band at the edge, frequently, but not always, has in the center a transparent protuberance, a sort of knob like that on the lid of a preserve-jar. Its entire surface is smooth and glossy, with no other ornament than its simplicity. The color varies according to the degree of maturity. When recently laid the eggs are of a uniform straw-yellow: later, owing to the gradual organization of the germ, they turn a pale orange, with a triangular bright-red patch in the center of the lid. When empty they are a magnificent, pellucid opal-white, except the lid, which has become transparent as glass.

Of the clutches of eggs obtained the most numerous was a patch of nine rows, each containing about a dozen eggs. The total was thus about a hundred. But usually the number of eggs is smaller than this, amounting to only half as many or less. Groups containing about a score of eggs are not uncommon. The enormous difference between these extremes testifies to multiple layings at different spots, which, in view of the insect's rapid flight, may be at quite a distance from one another. This detail will be of value when the time comes.

The Pale-Green Pentatoma (*P. praesinum*) molds her eggs in little barrels, ovoid at the bottom and adorned over their whole surface with a network of fine polygonal meshes in relief. Their color is a sooty brown, and, after the hatching, a very light brown. The largest groups of eggs contain thirty or so. It is probably to this species that the eggs belong which first attracted my attention on a sprig of asparagus.

As for the Berry Pentatoma (*P. baccarum*) here we again have barrels with rounded ends, covered all over the surface with a tracery of meshes. At first they are opaque and dark; then, being empty, they become translucent and white or pale-pink. Of these eggs I find groups of fifty and others of fifteen or even less.

That blessed plant of the kitchen-gardens, the cabbage, gives me the Ornate Pentatoma (*P. ornatum*), striped black and red. The eggs of this species are the prettiest of all in coloring. They are like little casks with the two ends convex, especially the lower. The microscope shows us a surface engraved with pits, like those of a thimble, arranged with exquisite regularity. At the top and bottom of the cylinder there is a broad dull-black band; on the sides is a wide white belt with four large black spots symmet-

rically placed. The lid, surrounded with snow-white filaments and edged with white, swells into a black dome with a central white spot. In short, a funeral urn, with its violent contrast of coal-black and creamy white. The Etruscans would have considered it a magnificent model for their burial vessels.

These eggs, with their funeral ornamentation, are arranged in small groups, generally in two rows. There are hardly a dozen all told: a fresh proof that the eggs must be laid in a number of batches and at different points; for the Cabbage Bug cannot limit herself to this paltry number when one of her relatives exceeds the hundred.

May is not over before the various batches of eggs collected and placed in tubes hatch out, first one and then another. Two or three weeks are enough to develop the germ. This is the time for constant vigilance, if I wish to understand the mechanism employed for the emergence and, above all, the function of the strange tool, with the three black arms, which I find in every shell, at the edge of the opening, once the new-born larva has departed.

Those eggs which are translucent from the outset – for example, those of the Black-horned Pentatoma – enable me, in the first place, to discover that the implement of unknown use makes its appearance rather late, when the approaching deliverance is announced by a change in the color of the lid. It is not, therefore, an original part of the egg, as this descended from the ovaries; it is elaborated during the process of development, and even at a somewhat advanced phase, when the little Bug has already been formed.

We must therefore cease to regard it, as I did at first, as a spring, a lock, some sort of a hinge to hold the lid in place. An actual device for keeping the egg closed and protecting the germ would have to be in existence when the egg was laid. And it is just at the end, when the time has come to leave it, that the egg reveals this device. It is a question no longer of closing, but of opening. And, in this case, might not the puzzling implement be a key, a lever to force open the lid, held on by thread-like rivets, and perhaps also by the glue of an adhesive? Assiduous patience will tell us.

Holding the magnifying-glass above my test-tubes, which I examine every moment, at last I witness the hatching. The process is just beginning. The lid is rising imperceptibly at one pole of its diameter; at the other it is tilting like a door on its hinges. The youngster has its back to the wall of the barrel, just below the edge of the lid, which is already gaping, a capital situation, enabling me to follow with some exactness the progress of the deliverance.

The little Bug, shrunken and motionless, has its head crowned with a skin cap, suspected rather than seen, so fine is it. Later, when it falls off, this cap will be plainly visible. It serves as the base of a trihedral angle. The three arms forming this angle are rigid and intensely black and look as if they ought to be of a horny nature. Two of them extend between the eyes, which are bright red; the third passes down behind the head and is connected with the others, right and left, by a dark, very fine line. I might very well regard these dark lines as tense threads, ligaments which brace the three arms of the apparatus and prevent them from slipping farther apart, thereby blunting the point of the angle, which is itself the key of the casket, that is, the rammer for pushing back the lid. This three-cornered miter protects the head, which is still soft and fleshy and incapable of forcing the obstruction: with its adamantine point truly applied right at the edge of the lid it has a firm grip of the disk which has to be unfastened.

This mechanism, this cap surmounted by an armored point, must have its motive force. Where is it? It is at the top of the head. Look carefully, and there, involving a certain small area, almost a point, you will see rapid pulsations, we might almost say piston-strokes, produced, beyond a doubt, by sudden waves of blood. By hurriedly injecting what little fluid its body contains under its pliant cranium, the tiny creature turns its weakness into energy. The three-cornered helmet rises, pushing upwards, always pressing its point firmly on the same point of the lid. No blow is struck upon the tool; there is no intermittent percussion, but a continuous thrust.

The operation is so laborious that it lasts for more than an hour. By imperceptible degrees the lid is unfastened and rises obliquely, but as a rule continues to adhere to the rim of the vase at the opposite pole of the diameter. At this pivotal point, where it would seem that there must be a hinge, the lens reveals nothing peculiar. Here, as every elsewhere, there is a mere row of threads, drawn down to form rivets for closing the cask. On the side opposite the point attacked, these rivets, less disturbed than the rest, do not quite give way, act as a hinge.

Little by little the tiny creature emerges from its shell. The legs and antennæ, economically folded over the thorax and abdomen, are completely motionless. Nothing moves, yet the Bug protrudes farther and farther from its casket, doubtless with the aid of a process like that employed by the larva of the Balaninus,[2] on leaving its nut. The flow of blood which causes the piston-strokes of the cranium distends also that part of the body which is already free and converts it into a supporting cushion; the hinder part, which is still imprisoned, is diminished accordingly and in its turn enters the narrow opening. The insect passes through a draw-plate, so gently and carefully that the most I can detect is a tentative rocking to and fro at distant intervals as it drags itself from its socket.

At last the rivets are forced, the casket is open, and the lid, now on a slant, is sufficiently raised. The three-cornered miter has done its work. What will become of it? Henceforth useless as a tool, it has to disappear; and, as a matter of fact, I see it discarded. The filmy head-dress which served as its foundation tears, becomes a tattered rag and very slowly slips over the Bug's ventral surface, dragging with it the hard little black contrivance, which still retains its shape. Scarcely has this relic slipped mid-way down the belly when the tiny creature, hitherto motionless in the attitude of a mummy, frees its legs and antennæ from their economical position, stretches them out and impatiently waves them to and fro. It is over: the insect leaves its sheath.

The instrument of release, still in the shape of a T with arms bent slightly downwards and sideways, remains sticking to the wall of the shell, near the opening. Long after the insect's departure the lens finds the ingenious triangle in its place. Its formation is the same in the various Pentatomæ; but until we surprise the insect in the act of hatching its function is incomprehensible.

A word more on the manner of opening the lidded casket. I have said that the young Bug has its back to the wall of the little barrel, as far as possible from the center. It is here that it is born, dons its tiara and afterwards pushes with its head. Why does it not occupy the central region, a position which would seem to be prescribed by the shape of the egg and the more effectual protection of the grub's early frailty? Can there be any advantage in being born elsewhere, on the very circumference?

Yes, there is, and a very distinct advantage, of a mechanical order. With the top of its head, which throbs with the rushes of blood, the new-born insect thrusts his pointed cap against the lid to be unfastened. What can be the cranial thrust of a drop of albumen but lately congealed into a living entity? He would be a bold man who should venture to reply, so far is it beneath all evaluation. And this mere nothing has to push open the solid lid of the box.

Let us picture the thrust applied to the center. In that case the effort to dislodge the lid, the veriest trifle of an effort, would be uniformly distributed over the entire circumference, and all the rivets which fasten it would play their part in the resistance offered. Singly, the stitches would give way before the tiny force available; but all together they are invincible. The method of the central thrust is therefore impracticable.

If we wished to loosen a nailed plank, it would be an illogical action to bang it in the middle. The whole of the nails would react in a common and insurmountable resistance. On the contrary, we attack it at one end; we apply the leverage of our implement progressively to one nail after another. The little Bug in its casket does much the same: it pushes out the extreme edge of the lid, so that, beginning at the point attacked, the rivets give way, one by one. The total resistance is overcome because it is divided.

Well done, little Bug! You have your own science of mechanics, based on the same laws as ours; you know the secrets of the lever and the lifting-jack. To break its shell, the nascent bird grows a callosity on its beak, a pick-axe point whose function is to break down the chalky wall piecemeal. When the task is accomplished this callus, the tool of a day, disappears. You have something better than the bird's device.

When the hour of your emergence comes, you don a cap in which three stiff ribs converge to a point. At the base of this appliance your soft cranium acts like the piston of a hydraulic press. Thus attacked, the roof of your hut is unfastened and thrown back. The bird's callosity disappears when the shell is in pieces; so does the miter with which you push out the head of your barrel. As soon as the lid opens wide enough to let you pass, you doff your cap with its tripod of rods.

Your egg, however, is not broken; there is no violent demolition such as that practiced by the bird. When empty, the eggshell is not a ruin: it is still the graceful little egg that it was in the beginning, rendered yet more exquisite by its translucence, which enhances its beauties. In what school, little Bug, did you learn the art of opening your natal casket and the use of your little contrivance? There are those who will say:

"In the school of chance."

But you, in all humility, cock your miter and reply:

"That's not true."

The Pentatoma is noted for another detail, which, if it were definitely proved, would surpass a hundredfold the marvels of the egg. I quote the following passage from de Geer,[3] the Swedish Réaumur:[4]

"The Bugs of this species (*Pentatoma griseum*) live on the birch-tree. In the early part of July, I found several of them accompanied by their young. Each mother was surrounded by a troop of young ones, to the number of twenty, thirty and even forty. She always kept close beside them, commonly on one of the catkins of the tree that contained her eggs, and sometimes on a leaf. I have noted that these little Bugs and their mother do not always remain on the same spot, and that as soon as the mother begins to move away all her little ones follow her, stopping whenever the mother calls a halt. She thus leads them from catkin to catkin or leaf to leaf and takes them wherever she pleases, as a Hen does her Chicks.

"There are Bugs that do not leave their offspring; they even keep watch

over them and take the greatest care of them while they are young. One day I happened to cut a young birch-branch peopled with such a family and I first observed the extremely uneasy mother, incessantly beating her wings with a rapid movement, without, however, stirring from the spot, as though to drive away the enemy that had just approached, whereas, in any other circumstances, she would at once have flown away or sought to escape, which proves that she was remaining only to defend her young."

M. Karl de Geer has observed that it is chiefly against the male of her species that the mother Bug is obliged to defend her young, because he tries to devour them wherever he comes upon them; and on such occasions she always tries with all her might to protect them against his attacks.

In his *Curiosités d'historie naturelle*, Boitard still further embellishes the picture of family life painted by de Geer:

"It is most curious," he says, "to see how the mother Bug, when a few drops of rain are falling, leads her young under a leaf or the fork of a branch to shelter them. Even there her anxious affection is not reassured; she drives them into a closely-packed flock, places herself in their midst and covers them with her wings, which she spreads over them umbrella-wise; and, in spite of the discomfort of her position, she retains this attitude of a brooding Hen until the storm has blown over."

Shall I confess it? This umbrella made of the mother's wings during showery weather, this procession of a Hen leading her Chicks, this devotion in warding off the attacks of a father inclined to devour his family leave me just a little incredulous, without surprising me, experience having taught me that the books are full of little anecdotes incapable of surviving the ordeal of a strict investigation.

An incomplete observation, wrongly interpreted, sets the story going. Then come the compilers, who faithfully hand down the legend, the unsound fruit of the imagination; and error, confirmed by repetition, becomes an article of faith. What, for example, was not reported of the Sacred Beetle and her pill, the Necrophorus[5] and her work of burial, the Hunting Wasp and her game, the Cicada and her well, before the truth was arrived at? The real, which is perfectly simple, and supremely beautiful, too often escapes us, giving way before the imaginary, which is less troublesome to acquire. Instead of going back to the facts and seeing for ourselves, we blindly follow tradition. To-day no one would write a few lines on the Pentatomæ without dragging in the Swedish naturalist's doubtful story, and no one, as far as I know, has mentioned the genuine marvels connected with the mechanism of the hatching.

What can de Geer have seen? The observer's high standing gives us confidence; none the less, I shall take the liberty of experimenting in my turn before accepting the master's statements.

The Grey Bug, the subject of my story, is less frequent than the others in my neighborhood: on the rosemaries in the enclosure, my field of exploration, I find three or four which, when placed under glass, do not give me any eggs. The set-back does not seem irreparable: what the grey refuses to reveal the green or the yellow or the red-and-black striped – one and all of similar formation and like habits – will show me. In species so closely akin, the family cares of the one must, in all but a few details, be re-produced in the others. Let us then note how the four Pentatomæ reared in captivity behave in the matter of their new-born young. Their unani-mous testimony will convince us.

At the very outset I was struck by a fact which disagreed with what I had a right to expect in a future Hen leading her Chicks. The mother pays no attention to her eggs. When the last has been laid in its place at the ex-treme end of the last row, she makes off, heedless of what she has left be-hind her. She does not trouble about it any more, does not return to it. If the hazards of her wanderings lead her up to it, she steps on the heap, crosses it and passes on, indifferent. The evidence leaves nothing to be de-sired: the coming upon a patch of eggs is an incident of no interest to the mother.

We must not attribute this negligence to the aberrations which may possibly occur in a state of captivity. In the perfect liberty of the fields I have come across many batches of eggs, perhaps including those of the Grey Bug; never have I seen the mother standing by her eggs, which she would have to do if her family required protection as soon as hatched.

The gravid mother is a quick flier and of a vagabond temperament. Once she has flown to a considerable distance from the leaf which has re-ceived her eggs, how is she to remember, two or three weeks later, that the hour for hatching is at hand? How is she to find her eggs again? Moreover, how is she to distinguish them from those of another mother? To believe her capable of such feats of clairvoyance and memory in the immensity of the open fields would be midsummer madness.

Never, I say, did I detect a mother permanently posted beside the eggs which she had fastened to a leaf. Further, the total emission is split up into partial deposits dispersed at random, so that the whole tribe comprises a series of clans encamped here and there, often removed to considerable distances which it is impossible to specify.

To rediscover these flocks at the time of the hatching, which falls earlier or later according to the date of production and the degree of exposure to the sun; then, from all over the country-side, to gather into one herd the whole of her very frail and short-legged offspring: this were an obvious impossibility. Let us nevertheless suppose that, by a stroke of good fortune, one of these groups is found and recognized and that the mother devotes herself to it. The others are necessarily abandoned. They thrive none the less well for that. Why, then, should some of the young Bugs be so strangely favored by maternal solicitude while the majority are able to do without it? Such peculiarities make one suspicious.

De Geer speaks of groups of twenty. These, we are forced to believe, were not the complete family, but detachments sprung from a partial laying. A Pentatoma smaller than the Grey Bug has given me, in one single deposit, more than a hundred eggs. This fecundity must be the general rule where the mode of life is the same. Apart from the twenty watched, then, what became of the rest, left to their own devices?

With all due respect to the Swedish naturalist, the tender cares of the mother Bug and the unnatural appetites of the father eating his children must be relegated to the fairy-tales with which history is crammed. I can obtain, in my breeding-cages, as many hatchings as I wish. The parents are

close at hand, under the same cover. What do they do respectively in the presence of the little ones?

Nothing whatever: the fathers do not hasten to slaughter their progeny nor do the mothers hasten to their rescue. They wander to and fro on the wire trellis; they take their rest in the restaurant provided by a tuft of rosemary; they pass through the groups of new-born Bugs and topple them over, without evil intent, but also without the least consideration. They are so small, the poor little wretches, and so feeble! A passer-by who grazes them with the tip of his foot turns them over on their backs. Like overturned Tortoises, they vainly kick and wriggle; no one heeds them.

Come then, O devoted mother! Since your family is beset by the danger of capsizing and other disagreeable accidents, place yourself at their head; lead them, step by step, into peaceful pastures; cover them with the buckler of your wing-cases! Any one waiting to observe these beautiful actions, these admirable and edifying moral characteristics, will waste his time and his patience. In three months of diligent watching I never saw, on the part of my charges, any action which in any way suggested the maternal solicitude so often extolled by the compilers of history.

Nature the universal nurse, *alma parens rerum*, is infinitely tender in her treatment of the germs, the treasure of the future; she is a harsh step-mother to the parent. As soon as the creature is capable of supporting itself, she delivers it without pity to life's cruel schooling, which teaches it to resist in the fierce struggle for existence. At first a tender mother, she gives the Pentatoma a delightful casket with a sealed lid to guard the budding flesh from harm; she caps the tiny insect with a mechanical device to set it free, a masterpiece of delicate ingenuity; and then, a stern schoolmistress, she says to the little one:

"I am leaving you. You must now fend for yourself in the hurly-burly of the world."

And the little insect does fend for itself. I see the new-born Bugs, pressed close against one another, remaining for some days on the patch of empty egg-shells. Their flesh grows firmer and their coloring brighter. Mothers pass at no great distance: none of them pays any attention to the drowsy company.

When hunger comes, one of the little ones moves away from the group in search of a canteen; the others follow; they love to feel shoulder touching shoulder, like grazing Sheep. The first to move draws the whole band after him; they make their way in a flock to the tender spots where they insert their suckers and drink their fill; whereupon all return to their native village, seeking a resting-place on the tops of the empty eggs. These expeditions in common are repeated within an increasing radius, till at last, having grown a little stronger, the community, becoming emancipated, makes off and disperses, no longer returning to the place of its birth. Henceforth each lives as he pleases.

What would happen if, when the flock is moving about, a mother were encountered, slow-stepping as the sober Bugs so often are? The little ones, I fancy, would confidently follow their chance-met leader, as they follow those among themselves who are the first to make a start. We should then see something like the Hen at the head of her Chicks; accident would give all the appearance of maternal solicitude to a stranger quite indifferent to the mob of children at her heels.

The worthy de Geer, it seems to me, must have been deceived by such meetings as these, in which maternal care played no part whatever. A little coloring, by way of involuntary adornment, completed the picture; and since then the domestic virtues of the Grey Bug have been lauded in all the books.

Curlew

NOTES

1. See *The Glow-worm and Other Beetles*: chaps. xviii and xix. — *Translator's Note.*

2. For the Nut-weevil, see *The Life of the Weevil*, by J. Henri Fabre, translated by Alexander Teixeira de Mattos: chap. vi; also his *Social Life in the Insect World*, translated by Bernard Miall. — *Translator's Note.*

3. Baron Karl de Geer (1720–1778), author of *Mémoires pour servir à l'histoire des insectes.* — *Translator's Note.*

4. René Antoine Ferchault de Réaumur (1683–1757), author of *Mémoires pour servir à l'histoire naturelle des insectes* and inventor of the Réaumur thermometer-scale. — *Translator's Note.*

5. Or Burying-beetle. See *The Glow-worm and Other Beetles:* chaps. xi and xii. — *Translator's Note.*

Megachile centuncularis

THE HALICTI: THE PORTRESS

Leaving our village is no very serious matter when we are children. We even look on it as a sort of holiday. We are going to see something new, those magic pictures of our dreams. With age come regrets; and the close of life is spent in stirring up old memories. Then the beloved village reappears, in the biograph of the mind, embellished, transfigured by the glow of those first impressions; and the mental image, superior to the reality, stands out in amazingly clear relief. The past, the far-off past, was only yesterday; we see it, we touch it.

For my part, after three-quarters of a century, I could walk with my eyes closed straight to the flat stone where I first heard the soft chiming note of the Midwife Toad; yes, I should find it to a certainty, if time, which devastates all things, even the homes of Toads, has not moved it or perhaps left it in ruins.

I see, on the margin of the brook, the exact position of the alder-trees whose tangled roots, deep under the water, were a refuge for the Crayfish. I should say:

Goldfinch

"It is just at the foot of that tree that I had the unutterable bliss of catching a beauty. She had horns so long . . . and enormous claws, full of meat, for I got her just at the right time."

I should go without faltering to the ash under whose shade my heart beat so loudly one sunny spring morning. I had caught sight of a sort of white, cottony ball among the branches. Peeping from the depths of the wadding was an anxious little head with a red hood to it. O what unparalleled luck! It was a Goldfinch, sitting on her eggs.

Compared with a find like this, lesser events do not count. Let us leave them. In any case, they pale before the memory of the paternal garden, a tiny hanging garden of some thirty paces by ten, situated right at the top of the village. The only spot that overlooks it is a little esplanade on which stands the old castle[1] with the four turrets that have now become dovecotes. A steep path takes you up to this open space. From my house on, it is more like a precipice than a slope. Gardens buttressed by walls are staged in terraces on the sides of the funnel-shaped valley. Ours is the highest; it is also the smallest.

There are no trees. Even a solitary apple-tree would crowd it. There is a patch of cabbages, with a border of sorrel, a patch of turnips and another of lettuces. That is all we have in the way of garden-stuff; there is no room for more. Against the upper supporting-wall, facing due south, is a vine-arbor which, at intervals, when the sun is generous, provides half a basketful of white muscatel grapes. These are a luxury of our own, greatly envied by the neighbors, for the vine is unknown outside this corner, the warmest in the village.

A hedge of currant-bushes, the only safeguard against a terrible fall, forms a parapet above the next terrace. When our parents' watchful eyes are off us, we lie flat on our stomachs, my brother and I, and look into the abyss at the foot of the wall bulging under the thrust of the land. It is the garden of monsieur le notaire.

There are beds with box-borders in that garden; there are pear-trees reputed to give pears, real pears, more or less good to eat when they have ripened on the straw all through the late autumn. In our imagination, it is a spot of perpetual delight, a paradise, but a paradise seen the wrong way up: instead of contemplating it from below, we gaze at it from above. How happy they must be with so much space and all those pears!

We look at the hives, around which the hovering Bees make a sort of russet smoke. They stand under the shelter of a great hazel. The tree has sprung up all of itself in a fissure of the wall, almost on the level of our currant-bushes. While it spreads its mighty branches over the notary's hives, its roots, at least, are on our land. It belongs to us. The trouble is to gather the nuts.

I creep along astride the strong branches projecting horizontally into space. If I slip or if the support breaks, I shall come to grief in the midst of the angry Bees. I do not slip and the support does not break. With the

bent switch which my brother hands me, I bring the finest clusters within my reach. I soon fill my pockets. Moving backwards, still straddling my branch, I recover *terra firma*. O wondrous days of litheness and assurance, when, for a few filberts, on a perilous perch we braved the abyss!

Enough. These reminiscences, so dear to my dreams, do not interest the reader. Why stir up more of them? I am content to have brought this fact into prominence: the first glimmers of light penetrating into the dark chambers of the mind leave an indelible impression, which the years make fresher instead of dimmer.

Obscured by everyday worries, the present is much less familiar to us, in its petty details, than the past, with childhood's glow upon it. I see plainly in my memory what my prentice eyes saw; and I should never succeed in reproducing with the same accuracy what I saw last week. I know my village thoroughly, though I quitted it so long ago; and I know hardly anything of the towns to which the vicissitudes of life have brought me. An exquisitely sweet link binds us to our native soil; we are like the plant that has to be torn away from the spot where it put out its first roots. Poor though it be, I should love to see my own village again; I should like to leave my bones there.

Does the insect in its turn receive a lasting impression of its earliest visions? Has it pleasant memories of its first surroundings? We will not speak of the majority, a world of wandering gipsies who establish themselves anywhere provided that certain conditions be fulfilled; but the others, the settlers, living in groups: do they recall their native village? Have they, like ourselves, a special affection for the place which saw their birth?

Yes, indeed they have: they remember, they recognize the maternal abode, they come back to it, they restore it, they colonize it anew. Among many other instances, let us quote that of the Zebra Halictus. She will show us a splendid example of love for one's birthplace translating itself into deeds.

The Halictus' spring family acquire the adult form in a couple of months or so; they leave the cells about the end of June. What goes on inside these neophytes as they cross the threshold of the burrow for the first time? Something, apparently, that may be compared with our own impressions of childhood. An exact and indelible image is stamped on their virgin memories. Despite the years, I still see the stone whence came the resonant notes of the little Toads, the parapet of currant-bushes, the notary's garden of Eden. These trifles make the best part of my life. The Halictus sees in the same way the blade of grass whereon she rested in her first flight, the bit of gravel which her claw touched in her first climb to the top of the shaft. She knows her native abode by heart just as I know my village. The locality has become familiar to her in one glad, sunny morning.

Spadefoot toad

She flies off, seeks refreshment on the flowers near at hand and visits the fields where the coming harvests will be gathered. The distance does not lead her astray, so faithful are her impressions of her first trip; she finds the encampment of her tribe; among the burrows of the village, so numerous and so closely resembling one another, she knows her own. It is the house where she was born, the beloved house with its unforgettable memories.

But, on returning home, the Halictus is not the only mistress of the house. The dwelling dug by the solitary Bee in early spring remains, when summer comes, the joint inheritance of the members of the family. There are ten cells, or thereabouts, underground. Now from these cells there have issued none but females. This is the rule among the three species of Halicti that concern us now and probably also among many others, if not all. They have two generations in each year. The spring one consists of females only; the summer one comprises both males and females, in almost equal numbers.

The household, therefore, if not reduced by accidents, above all if not starved by the usurping Gnat, would consist of half-a-score of sisters, none but sisters, all equally industrious and all capable of procreating without a nuptial partner. On the other hand, the maternal dwelling is no hovel; far from it: the entrance-gallery, the principal room of the house, will serve quite well, after a few odds and ends of refuse have been swept away. This will be so much gained in time, ever precious to the Bee. The cells at the bottom, the clay cabins, are also nearly intact. To make use of them, it will be enough for the Halictus to polish up the stucco with her tongue.

Well, which of the survivors, all equally entitled to the succession, will inherit the house? There are six of them, seven, or more, according to the chances of mortality. To whose share will the maternal dwelling fall?

There is no quarrel between the interested parties. The mansion is recognized as common property without dispute. The sisters come and go

peacefully through the same door, attend to their business, pass and let the others pass. Down at the bottom of the pit, each has her little demesne, her group of cells dug at the cost of fresh toil, when the old ones, now insufficient in number, are occupied. In these recesses, which are private estates, each mother works by herself, jealous of her property and of her privacy. Every elsewhere, traffic is free to all.

The exits and entrances in the working fortress provide a spectacle of the highest interest. A harvester arrives from the fields, the feather-brushes of her legs powdered with pollen. If the door be open, the Bee at once dives underground. To tarry on the threshold would mean waste of time; and the business is urgent. Sometimes, several appear upon the scene at almost the same moment. The passage is too narrow for two, especially when they have to avoid any untimely contact that would make the floury burden fall to the floor. The nearest to the opening enters quickly. The others, drawn up on the threshold in the order of their arrival, respectful of one another's rights, await their turn. As soon as the first disappears, the second follows after her and is herself swiftly followed by the third and then the others, one by one.

Sometimes, again, there is a meeting between a Bee about to come out and a Bee about to go in. Then the latter draws back a little and makes way for the former. The politeness is reciprocal. I see some who, when on the point of emerging from the pit, go down again and leave the passage free for the one who has just arrived. Thanks to this mutual spirit of accommodation, the business of the house proceeds without impediment.

Let us keep our eyes open. There is something better than the well-preserved order of the entrances. When an Halictus appears, returning from her round of the flowers, we see a sort of trap-door, which closed the house, suddenly fall and give a free passage. As soon as the new arrival has entered, the trap rises back into its place, almost level with the ground, and closes the entrance anew. The same thing happens when the insects go out. At a request from within, the trap descends, the door opens and the Bee flies away. The outlet is closed forthwith.

What can this valve be which, descending or ascending in the cylinder of the pit, after the fashion of a piston, opens and closes the house at each departure and at each arrival? It is an Halictus, who has become portress of the establishment. With her large head, she makes an impassable barrier at the top of the entrance-hall. If any one belonging to the house wants to go in or out, she "pulls the cord," that is to say, she withdraws to a spot where the gallery becomes wider and leaves room for two. The other passes. She then at once returns to the orifice and blocks it with the top of her head. Motionless, ever on the look-out, she does not leave her post save to drive away importunate visitors.

Let us profit by her brief appearances outside to take a look at her. We recognize in her an Halictus similar to the others, which are now busy harvesting; but the top of her head is bald and her dress is dingy and threadbare. All the nap is gone; and one can hardly make out the handsome stripes of red and brown which she used to have. These tattered, workworn garments make things clear to us.

This Bee who mounts guard and performs the office of a portress at the entrance to the burrow is older than the others. She is the foundress of the establishment, the mother of the actual workers, the grandmother of the present grubs. In the springtime of her life, three months ago, she wore herself out in solitary labors. Now that her ovaries are dried up, she takes a well-earned rest. No, rest is hardly the word. She still works, she assists the household to the best of her power. Incapable of being a mother for a second time, she becomes a portress, opens the door to the members of her family and makes strangers keep their distance.

The suspicious Kid,[2] looking through the chink, said to the Wolf:

"Show me a white foot, or I shan't open the door."

No less suspicious, the grandmother says to each comer:

"Show me the yellow foot of an Halictus, or you won't be let in."

None is admitted to the dwelling unless she be recognized as a member of the family.

See for yourselves. Near the burrow passes an Ant, an unscrupulous adventuress, who would not be sorry to know the meaning of the honeyed fragrance that rises from the bottom of the cellar.

"Be off, or you'll catch it!" says the portress, wagging her neck.

As a rule the threat suffices. The Ant decamps. Should she insist, the watcher leaves her sentry-box, flings herself upon the saucy jade, buffets her and drives her away. The moment the punishment has been administered, she returns to her post.

Next comes the turn of a Leaf-cutter (*Megachile albocincta*, Pérez), which, unskilled in the art of burrowing, utilizes, after the manner of her kin, the old galleries dug by others. Those of the Zebra Halictus suit her very well, when the terrible Gnat has left them vacant for lack of heirs. Seeking for a home wherein to stack her robinia-leaf honey-pots, she often makes a flying inspection of my colonies of Halicti. A burrow seems to take her fancy; but, before she sets foot on earth, her buzzing is noticed by the sentry, who suddenly darts out and makes a few gestures on the threshold of her door. That is all. The Leaf-cutter has understood. She moves on.

Sometimes, the Megachile has time to alight and insert her head into the mouth of the pit. In a moment, the portress is there, comes a little higher and bars the way. Follows a not very serious contest. The stranger quickly recognizes the rights of the first occupant and, without insisting, goes to seek an abode elsewhere.

An accomplished marauder (*Cælioxys caudata*, Spin.), a parasite of the Megachile, receives a sound drubbing under my eyes. She thought, the feather-brain, that she was entering the Leaf-cutter's establishment! She soon finds out her mistake; she meets the door-keeping Halictus, who administers a sharp correction. She makes off at full speed. And so with the others which, through inadvertence or ambition, seek to enter the burrow.

The same intolerance exists among the different grandmothers. About the middle of July, when the animation of the colony is at its height, two sets of Halicti are easily distinguishable: the young mothers and the old. The former, much more numerous, brisk of movement and smartly arrayed, come and go unceasingly from the burrows to the fields and from the fields to the burrows. The latter, faded and dispirited, wander idly from hole to hole. They look as though they had lost their way and were incapable of finding their homes. Who are these vagabonds? I see in them afflicted ones bereft of a family through the act of the odious Gnat. Many burrows have been altogether exterminated. At the awakening of summer, the mother found herself alone. She left her empty house and went off in search of a dwelling where there were cradles to defend, a guard to mount.

But those fortunate nests already have their overseer, the foundress, who, jealous of her rights, gives her unemployed neighbor a cold reception. One sentry is enough; two would merely block the narrow guard-room.

I am privileged at times to witness a fight between two grandmothers. When the tramp in quest of employment appears outside the door, the lawful occupant does not move from her post, does not withdraw into the passage, as she would before an Halictus returning from the fields. Far from making way, she threatens the intruder with her feet and mandibles. The other retaliates and tries to force her way in notwithstanding. Blows are exchanged. The fray ends by the defeat of the stranger, who goes off to pick a quarrel elsewhere.

These little scenes afford us a glimpse of certain details of the highest interest in the habits of the Zebra Halictus. The mother who builds her nest in the spring no longer leaves her home, once her works are finished. Shut up at the bottom of the burrow, busied with the thousand cares of housekeeping, or else drowsing, she waits for her daughters to come out. When, in the summer heats, the life of the village recommences, having naught to do outside as a harvester, she stands sentry at the entrance to the hall, so as to let none in save the workers of the home, her own daughters. She wards off evilly-disposed visitors. None can enter without the door-keeper's consent.

There is nothing to tell us that the watcher ever deserts her post. Not once do I see her leave her house to go and seek some refreshment from the flowers. Her age and her sedentary occupation, which involves no great fatigue, perhaps relieve her of the need of nourishment. Perhaps, also, the young ones returning from their plundering may from time to time disgorge a drop of the contents of their crops for her benefit. Fed or unfed, the old one no longer goes out.

But what she does need is the joys of an active family. Many are deprived of these. The Gnat's burglary has destroyed the busy household.

The sorely-tried Bees abandon the deserted burrow. It is they who, ragged and careworn, wander through the village. When they move, their flight is only a short one; more often they remain motionless. It is they who, soured in their tempers, attack their fellows and seek to dislodge them. They grow rarer and more languid from day to day; then they disappear for good. What has become of them? The little Grey Lizard had his eye on them: they are easily snapped up.

Those settled in their own demesne, those who guard the honey-factory wherein their daughters, the heiresses of the maternal establishment, are at work, display wonderful vigilance. The more I see of them, the more I admire them. In the cool hours of the early morning, when the pollen-flour is not sufficiently ripened by the sun and while the harvesters are still indoors, I see them at their posts, at the top of the gallery. Here, motion-less, their heads flush with the earth, they bar the door to all invaders. If I look at them closely, they retreat a little and, in the shadow, await the indis-creet observer's departure.

I return when the harvesting is in full swing, between eight o'clock and twelve. There is now, as the Halicti go in or out, a succession of prompt withdrawals to open the door and of ascents to close it. The portress is in the full exercise of her functions.

In the afternoon, the heat is too great and the workers do not go to the fields. Retiring to the bottom of the house, they varnish the new cells, they make the round loaf that is to receive the egg. The grandmother is still up-stairs, stopping the door with her bald head. For her, there is no siesta dur-ing the stifling hours: the safety of the household requires her to forgo it.

I come back again at nightfall, or even later. By the light of a lantern, I again behold the overseer, as zealous and assiduous as in the day-time. The others are resting, but not she, for fear, apparently, of nocturnal dan-gers known to herself alone. Does she nevertheless end by descending to

the quiet of the floor below? It seems probable, so essential must rest be, after the fatigue
of such a vigil!

It is evident that, guarded in this manner,
the burrow is exempt from calamities similar
to those which, too often, depopulate it in
May. Let the Gnat come now, if she dare, to
steal the Halictus' loaves! Let her lie in wait as
long as she will! Neither her audacity nor her
slyness will make her escape the lynx eyes of
the sentinel, who will put her to flight with a
threatening gesture or, if she persist, crush her
with her nippers. She will not come; and we
know the reason: until spring returns, she is
underground in the pupa state.

But, in her absence, there is no lack, among
the Fly rabble, of other batteners on the toil of their fellow-insects. Whatever the job, whatever the plunder, you will find parasites there. And yet,
for all my daily visits, I never catch one of these in the neighborhood of
the summer burrows. How cleverly the rascals ply their trade! How well
aware are they of the guard who keeps watch at the Halictus' door! There
is no foul deed possible nowadays; and the result is that no Fly puts in an
appearance and the tribulations of last spring are not repeated.

The grandmother who, dispensed by age from maternal bothers, mounts
guard at the entrance of the home and watches over the safety of the family, tells us that in the genesis of the instincts sudden births occur; she
shows us the existence of a spontaneous aptitude which nothing, either in
her own past conduct or in the actions of her daughters, could have led us
to suspect. Timorous in her prime, in the month of May, when she lived

alone in the burrow of her making, she has become gifted, in her decline, with a superb contempt of danger and dares in her impotence what she never dared do in her strength.

Formerly, when her tyrant, the Gnat, entered the house in her presence, or, more often, stood face to face with her at the entrance, the silly Bee did not stir, did not even threaten the red-eyed bandit, the dwarf whose doom she could so easily have sealed. Was it terror on her part? No, for she attended to her duties with her usual punctiliousness; no, for the strong do not allow themselves to be thus paralyzed by the weak. It was ignorance of the danger, it was sheer fecklessness.

And behold, to-day, the ignoramus of three months ago knows the peril, knows it well, without serving any apprenticeship. Every stranger who appears is kept at a distance, without distinction of size or race. If the threatening gesture be not enough, the keeper sallies forth and flings herself upon the persistent one. Cowardice has developed into courage.

How has this change been brought about? I should like to picture the Halictus gaining wisdom from the misfortunes of the spring and capable thenceforth of looking out for danger; I would gladly credit her with having learnt in the stern school of experience the advantages of a patrol. I must give up the idea. If, by dint of gradual little acts of progress, the Bee has achieved the glorious invention of a janitress, how comes it that the fear of thieves is intermittent? It is true that, being by herself in May, she cannot stand permanently at her door: the business of the house takes precedence of everything else. But she ought, at any rate as soon as her offspring are victimized, to know the parasite and give chase when, at every moment, she finds her almost under her feet and even in her house. Yet she pays no attention to her.

The bitter experience of her ancestors, therefore, has bequeathed nothing to her of a nature to alter her placid character; nor have her own tribulations aught to do with the sudden awakening of her vigilance in July.

Like ourselves, animals have their joys and their sorrows. They eagerly make the most of the former; they fret but little about the latter, which, when all is said, is the best way of achieving a purely animal enjoyment of life. To mitigate these troubles and protect the progeny there is the inspiration of instinct, which is able without the counsels of experience to give the Halicti a portress.

When the victualling is finished, when the Halicti no longer sally forth on harvesting intent nor return all befloured with their spoils, the old Bee is still at her post, vigilant as ever. The final preparations for the brood are made below; the cells are closed. The door will be kept until everything is finished. Then grandmother and mothers leave the house. Exhausted by the performance of their duty, they go, somewhere or other, to die.

In September appears the second generation, comprising both males and females. I find both sexes wassailing on the flowers, especially the Compositæ, the centauries and thistles. They are not harvesting now: they are refreshing themselves, holding high holiday, teasing one another. It is the wedding-time. Yet another fortnight and the males will disappear, henceforth useless. The part of the idlers is played. Only the industrious ones remain, the impregnated females, who go through the winter and set to work in April.

I do not know their exact haunt during the inclement season. I expected them to return to their native burrow, an excellent dwelling for the winter, one would think. Excavations made in January showed me my mistake. The old homes are empty, are falling to pieces owing to the prolonged effect of the rains. The Zebra Halictus has something better than these muddy hovels: she has snug corners in the stone-heaps, hiding-places in the sunny walls and many other convenient habitations. And so the natives of a village become scattered far and wide.

In April, the scattered ones reassemble from all directions. On the well-flattened garden-paths a choice is made of the site for their common

labors. Operations soon begin. Close to the first who bores her shaft there is soon a second one busy with hers; a third arrives, followed by another and others yet, until the little mounds often touch one another, while at times they number as many as fifty on a surface of less than a square yard.

One would be inclined, at first sight, to say that these groups are accounted for by the insect's recollection of its birthplace, by the fact that the villagers, after dispersing during the winter, return to their hamlet. But it is not thus that things happen: the Halictus scorns to-day the place that once suited her. I never see her occupy the same patch of ground for two years in succession. Each spring she needs new quarters. And there are plenty of them.

Can this mustering of the Halicti be due to a wish to resume the old intercourse with their friends and relations? Do the natives of the same burrow, of the same hamlet, recognize one another? Are they inclined to do their work among themselves rather than in the company of strangers? There is nothing to prove it, nor is there anything to disprove it. Either for this reason or for others, the Halictus likes to keep with her neighbors.

This propensity is pretty frequent among peace-lovers, who, needing

little nourishment, have no cause to fear competition. The others, the big
eaters, take possession of estates, of hunting-grounds from which their
fellows are excluded. Ask a Wolf his opinion of a brother Wolf poaching
on his preserves. Man himself, the chief of consumers, makes for himself
frontiers armed with artillery; he sets up posts at the foot of which one
says to the other:

"Here's my side, there's yours. That's enough: now we'll pepper each
other."

And the rattle of the latest explosives ends the colloquy.

Happy are the peace-lovers. What do they gain by their mustering?
With them it is not a defensive system, a concerted effort to ward off the
common foe. The Halictus does not care about her neighbor's affairs. She
does not visit another's burrow; she does not allow others to visit hers.
She has her tribulations, which she endures alone; she is indifferent to the
tribulations of her kind. She stands aloof from the strife of her fellows.
Let each mind her own business and leave things at that.

But company has its attractions. He lives twice who watches the life of
others. Individual activity gains by the sight of the general activity; the animation of each one derives fresh warmth from the fire of the universal
animation. To see one's neighbors at work stimulates one's rivalry. And
work is the great delight, the real satisfaction that gives some value to life.
The Halictus knows this well and assembles in her numbers that she may
work all the better.

Sometimes she assembles in such multitudes and over such extents of
ground as to suggest our own colossal swarms. Babylon and Memphis,
Rome and Carthage, London and Paris, those frantic hives, occur to our
mind if we can manage to forget comparative dimensions and see a Cyclo-
pean pile in a pinch of earth.

It was in February. The almond-tree was in blossom. A sudden rush of
sap had given the tree new life; its boughs, all black and desolate, seem-

ingly dead, were becoming a glorious dome of snowy satin. I have always loved this magic of the awakening spring, this smile of the first flowers against the gloomy bareness of the bark.

And so I was walking across the fields, gazing at the almond-trees' carnival. Others were before me. An Osmia in a black velvet bodice and a red woollen skirt, the Horned Osmia, was visiting the flowers, dipping into each pink eye in search of a honeyed tear. A very small and very modestly-dressed Halictus, much busier and in far greater numbers, was flitting silently from blossom to blossom. Official science calls her *Halictus malachurus*, K. The pretty little Bee's godfather strikes me as ill-inspired. What has *malachurus*, calling attention to the softness of the rump, to do in this connection? The name of Early Halictus would better describe the almond-tree's little visitor.

None of the melliferous clan, in my neighborhood at least, is stirring as early as she is. She digs her burrows in February, an inclement month, subject to sudden returns of frost. When none as yet, even among her near kinswomen, dares to sally forth from winter-quarters, she pluckily goes to work, shine the sun ever so little. Like the Zebra Halictus, she has two generations a year, one in spring and one in summer; like her, too, she settles by preference in the hard ruts of the country roads.

Her mole-hills, those humble mounds any two of which would go easily into a Hen's egg, rise innumerous in my path, the path by the almond-trees which is the happy hunting-ground of my curiosity to-day. This path is a ribbon of road three paces wide, worn into ruts by the Mule's hoofs and the wheels of the farm-carts. A coppice of holm-oaks shelters it from the north wind. In this Eden with its well-caked soil, its warmth and quiet, the little Halictus has multiplied her mole-hills to such a degree that I cannot take a step without crushing some of them. The accident is not serious: the miner, safe underground, will be able to scramble up the crumbling sides of the mine and repair the threshold of the trampled home.

I make a point of measuring the density of the population. I count from forty to sixty mole-hills on a surface of one square yard. The encampment is three paces wide and stretches over nearly three-quarters of a mile. How many Halicti are there in this Babylon? I do not venture to make the calculation.

Speaking of the Zebra Halictus, I used the words *hamlet, village, township;* and the expressions were appropriate. Here the term *city* hardly meets the case. And what reason can we allege for these innumerable clusters? I can see but one: the charm of living together, which is the origin of society. Like mingles with like, without the rendering of any mutual service; and this is enough to summon the Early Halictus to the same way-side, even as the Herring and the Sardine assemble in the same waters.

NOTES

1. The Château de Saint-Léons standing just outside and above the village of Saint-Léons, where the author was born in 1823. See *The Life of the Fly:* chaps. vi and vii. – *Translator's Note.*

2. In La Fontaine's fable, *Le Loup, la Chèvre et le Chevreau.* – *Translator's Note.*

THE POND

The pond, the delight of my early childhood, is still a sight whereof my old eyes never tire. What animation in that verdant world! On the warm mud of the edges, the Frog's little Tadpole basks and frisks in its black legions; down in the water, the orange-bellied Newt steers his way slowly with the broad rudder of his flat tail; among the reeds are stationed the flotillas of Caddis-worms half-protruding from their tubes, which are now a tiny bit of stick and again a turret of little shells.

In the deep places, the Water-beetle dives, carrying with him his reserves of breath: an air-bubble at the tip of the wing-cases and, under the chest, a film of gas that gleams like a silver breastplate; on the surface, the ballet of those shimmering pearls, the Whirligigs, turns and twists about; hard by there skims the insubmersible troop of the Pond-skaters, who glide along with side-strokes similar to those which the cobbler makes when sewing.

Here are the Water-boatmen, who swim on their backs with two oars spread cross-wise, and the flat Water-scorpions; here, squalidly clad in mud, is the grub of the largest of our Dragon-flies, so curious because of its manner of progression: it fills its hinder-parts, a yawning funnel, with water, spurts it out again and advances just so far as the recoil of its hydraulic cannon.

Lymnæa
stagnalis

Planorbarius
corneus

The Molluscs abound, a peaceful tribe. At the bottom, the plump River-snails discreetly raise their lid, opening ever so little the shutters of their dwelling; on the level of the water, in the glades of the aquatic garden, the Pond-snails – Physa, Limnæa and Planorbis – take the air. Dark Leeches writhe upon their prey, a chunk of Earth-worm; thousands of tiny, reddish grubs, future Mosquitoes, go spinning around and twist and curve like so many graceful Dolphins.

Yes, a stagnant pool, though but a few feet wide, hatched by the sun, is an immense world, an inexhaustible mine of observation to the studious man and a marvel to the child who, tired of his paper boat, diverts his eyes and thoughts a little with what is happening in the water. Let me tell what I remember of my first pond, at a time when ideas began to dawn in my seven-year-old brain.

How shall a man earn his living in my poor native village, with its inclement weather and its niggardly soil? The owner of a few acres of grazing-land rears sheep. In the best parts, he scrapes the soil with the swing-plough; he flattens it into terraces banked by walls of broken stones. Pannierfuls of dung are carried up on donkey-back from the cow-shed. Then, in due season, comes the excellent potato, which, boiled and served hot in a basket of plaited straw, is the chief stand-by in winter.

Should the crop exceed the needs of the household, the surplus goes to feed a pig, that precious beast, a treasure of bacon and ham. The ewes supply butter and curds; the garden boasts cabbages, turnips and even a few hives in a sheltered corner. With wealth like that one can look fate in the face.

But we, we have nothing, nothing but the little house inherited by my mother and its adjoining patch of garden. The meager resources of the family are coming to an end. It is time to see to it and that quickly. What is to be done? That is the stern question which father and mother sat de-bating one evening.

Hop-o'-my-Thumb, hiding under the wood-cutter's stool, lis-tened to his parents overcome by want. I also, pretending to sleep, with my elbows on the table, listen not to blood-curdling de-signs, but to grand plans that set my heart rejoicing. This is how the matter stands: at the bottom of the village, near the church, at the spot where the water of the large roofed spring escapes from its under-ground weir and joins the brook in the valley, an enterpris-ing man, back from the war,[1] has set up a small tallow-factory. He sells the scrapings of his pans, the burnt fat, reeking of candle-grease, at a low price. He pro-claims these wares to be excellent for fattening ducks.

"Suppose we bred some ducks," says mother. "They sell very well in town. Henri would mind them and take them down to the brook."

"Very well," says father, "let's breed some ducks. There may be difficul-ties in the way; but we'll have a try."

That night, I had dreams of paradise: I was with my ducklings, clad in their yellow suits; I took them to the pond, I watched them have their bath, I brought them back again, carrying the more tired ones in a basket.

A month or two after, the little birds of my dreams were a reality. There were twenty-four of them. They had been hatched by two hens, of whom one, the big, black one, was an inmate of the house, while the other was borrowed from a neighbor.

To bring them up, the former is sufficient, so careful is she of her adopted family. At first, everything goes perfectly: a tub with two fingers' depth of water serves as a pond. On sunny days, the ducklings bathe in it under the anxious eye of the hen.

A fortnight later, the tub is no longer enough. It contains neither cresses crammed with tiny Shellfish nor Worms and Tadpoles, dainty morsels both. The time has come for dives and hunts amid the tangle of the water-weeds; and for us the day of trouble has also come. True, the miller, down by the brook, has fine ducks, easy and cheap to bring up; the tallow-smelter, who has extolled his burnt fat so loudly, has some as well, for he has the advantage of the waste water from the spring at the bottom of the village; but how are we, right up there, at the top, to procure aquatic sports for our broods? In summer, we have hardly water to drink!

Near the house, in a freestone recess, a scanty source trickles into a basin made in the rock. Four or five families have, like ourselves, to draw their water there with copper pails. By the time that the schoolmaster's donkey has slaked her thirst and the neighbors have taken their provision for the day, the basin is dry. We have to wait for four-and-twenty hours for it to fill. No, this is not the hole in which the ducks would delight nor indeed in which they would be tolerated.

There remains the brook. To go down to it with the troop of ducklings is fraught with danger. On the way through the village, we might meet cats, bold ravishers of small poultry; some surly mongrel might frighten and scatter the little band; and it would be a hard puzzle to collect it in its entirety. We must avoid the traffic and take refuge in peaceful and sequestered spots.

On the hills, the path that climbs behind the château[2] soon takes a sudden turn and widens into a small plain beside the meadows. It skirts a rocky slope whence trickles, level with the ground, a streamlet, forming a pond of some size. Here profound solitude reigns all day long. The ducklings will be well off; and the journey can be made in peace by a deserted foot-path.

You, little man, shall take them to that delectable spot. What a day it was that marked my first appearance as a herdsman of ducks! Why must there be a jar to the even tenor of such joys! The too-frequent encounter of my tender skin with the hard ground had given me a large and painful blister on the heel. Had I wanted to put on the shoes stowed away in the cupboard for Sundays and holidays, I could not. There was nothing for it but to go barefoot over the broken stones, dragging my leg and carrying high the injured heel.

Let us make a start, hobbling along, switch in hand, behind the ducks. They too, poor little things, have sensitive soles to their feet; they limp, they quack with fatigue. They would refuse to go any farther, if I did not, from time to time, call a halt under the shelter of an ash.

We are there at last. The place could not be better for my birdlets: shallow, tepid water, interspersed with muddy knolls and green eyots. The diversions of the bath begin forthwith. The ducklings clap their beaks and rummage here, there and everywhere; they sift each mouthful, rejecting the clear water and retaining the good bits. In the deeper parts, they point their sterns into the air and stick their heads under water. They are happy; and it is a blessed thing to see them at work. We will let them be. It is my turn to enjoy the pond.

What is this? On the mud lie some loose, knotted, soot-colored cords. One could take them for threads of wool like those which you pull out of an old ravelly stocking. Can some shepherdess, knitting a black sock and finding her work turn out badly, have begun all over again and, in her impatience, have thrown down the wool with all the dropped stitches? It really looks like it.

I take up one of those cords in my hand. It is sticky and extremely slack; the thing slips through the fingers before they can catch hold of it. A few of the knots burst and shed their contents. What comes out is a black globule, the size of a pin's head, followed by a flat tail. I recognize, on a very small scale, a familiar object: the Tadpole, the Frog's baby. I have seen enough. Let us leave the knotted cords alone.

The next creatures please me better. They spin round on the surface of the water and their black backs gleam in the sun. If I lift a hand to seize them, that moment they disappear, I know not where. It's a pity: I should have much liked to see them closer and to make them wriggle in a little bowl which I should have put ready for them.

Let us look at the bottom of the water, pulling aside those bunches of green string whence beads of air are rising and gathering into foam. There is something of everything underneath. I see pretty shells with compact whorls, flat as beans; I notice little worms carrying tufts and feathers; I make out some with flabby fins constantly flapping on their backs. What are they all doing there? What are their names? I do not know. And I stare at them for ever so long, held by the incomprehensible mystery of the waters.

At the place where the pond dribbles into the adjoining field are some alder-trees; and here I make a glorious find. It is a Scarab – not a very large one, oh no! He is smaller than a cherry-stone, but of an unutterable blue.

The angels in paradise must wear dresses of that 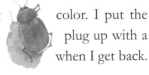 color. I put the glorious one inside an empty snail-shell, which I plug up with a leaf. I shall admire that living jewel at my leisure, when I get back. Other distractions summon me away.

The spring that feeds the pond trickles from the rock, cold and clear. The water first collects into a cup, the size of the hollow of one's two hands, and then runs over in a stream. These falls call for a mill: that goes without saying. Two bits of straw, artistically crossed upon an axis, provide the machinery; some flat stones set on edge afford supports. It is a great success: the mill turns admirably. My triumph would be complete, could I but share it. For want of other playmates, I invite the ducks.

Everything palls in this poor world of ours, even a mill made of two straws. Let us think of something else: let us contrive a dam to hold back the waters and form a pool. There is no lack of stones for the brickwork. I pick the most suitable; I break the larger ones. And, while collecting these blocks, suddenly I forget all about the dam which I meant to build.

On one of the broken stones, in a cavity large enough for me to put my fist in, something gleams like glass. The hollow is lined with facets gathered in sixes which flash and glitter in the sun. I have seen something like this in church, on the great saints'-days, when the light of the candles in the big chandelier kindles the stars in its hanging crystal.

We children, lying, in summer, on the straw of the threshing-floor, have told one another stories of the treasures which a dragon guards underground. Those treasures now return to my mind: the names of precious stones ring out uncertainly but gloriously in my memory. I think of the king's crown, of the princesses' necklaces. In breaking stones, can I have found, but on a much richer scale, the thing that shines quite small in my mother's ring? I want more such.

The dragon of the subterranean treasures treats me generously. He gives me his diamonds in such quantities that soon I possess a heap of

Calopteryx virgo

Sympetrum sanguineum

broken stones sparkling with magnificent clusters. He does more: he gives me his gold. The trickle of water from the rock falls on a bed of fine sand which it swirls into bubbles. If I bent over towards the light, I see something like gold-filings whirling where the fall touches the bottom. Is it really the famous metal of which twenty-franc pieces, so rare with us at home, are made? One would think so, from the glitter.

I take a pinch of sand and place it in my palm. The brilliant particles are numerous, but so small that I have to pick them up with a straw moistened in my mouth. Let us drop this: they are too tiny and too bothersome to collect. The big, valuable lumps must be farther on, in the thickness of the rock. We'll come back later; we'll blast the mountain.

I break more stones. Oh, what a queer thing has just come loose, all in one piece! It is turned spiral-wise, like certain flat Snails that come out of the cracks of old walls in rainy weather. With its gnarled sides, it looks like a little ram's-horn. Shell or horn, it is very curious. How do things like that find their way into the stone?

Treasures and curiosities make my pockets bulge with pebbles. It is late and the little ducklings have had all they want to eat. Come along, youngsters, let's go home. My blistered heel is forgotten in my excitement.

The walk back is a delight. A voice sings in my ear, an untranslatable voice, softer than any language and bewildering as a dream. It speaks to me for the first time of the mysteries of the pond; it glorifies the heavenly insect which I hear moving in the empty snail-shell, its temporary cage; it whispers the secrets of the rock, the gold-filings, the faceted jewels, the ram's-horn turned to stone.

Poor simpleton, smother your joy! I arrive. My parents catch sight of my bulging pockets, with their disgraceful load of stones. The cloth has given way under the rough and heavy burden.

"You rascal!" says father, at sight of the damage. "I send you to mind the ducks and you amuse yourself picking up stones, as though there weren't enough of them all round the house! Make haste and throw them away!"

Broken-hearted, I obey. Diamond, gold-dust, petrified ram's-horn, heavenly Beetle are all flung on a rubbish-heap outside the door.

Mother bewails her lot:

"A nice thing, bringing up children to see them turn out so badly! You'll bring me to my grave. Green stuff I don't mind: it does for the rabbits. But stones, which ruin your pockets; poisonous animals, which'll sting your hand: what good are they to you, silly? There's no doubt about it: some one has thrown a spell over you!"

Yes, my poor mother, you were right, in your simplicity: a spell had been cast upon me; I admit it to-day. When it is hard enough to earn one's bit of bread, does not improving one's mind but render one more meet for suffering? Of what avail is the torment of learning to the derelicts of life?

A deal better off am I, at this late hour, dogged by poverty and knowing that the diamonds of the duck-pool were rock-crystal, the gold-dust mica, the stone horn an Ammonite and the sky-blue Beetle a Hoplia! We poor men would do better to mistrust the joys of knowledge: let us dig our furrow in the fields of the commonplace, flee the temptations of the pond, mind our ducks and leave to others, more favored by fortune, the job of explaining the world's mechanism, if the spirit moves them.

And yet no! Alone among living creatures, man has the thirst for knowledge; he alone pries into the mysteries of things. The least among us will utter his whys and his wherefores, a fine pain unknown to the brute beast. If these questionings come from us with greater persistence, with a more imperious authority, if they divert us from the quest of lucre, life's only object in the eyes of most men, does it become us to complain? Let us be careful not to do so, for that would be denying the best of all our gifts.

Let us strive, on the contrary, within the measure of our capacity, to force a gleam of light from the vast unknown; let us examine and question and, here and there, wrest a few shreds of truth. We shall sink under the task; in the present ill-ordered state of society, we shall end, perhaps, in the workhouse. Let us go ahead for all that: our consolation shall be that we have increased by one atom the general mass of knowledge, the incomparable treasure of mankind.

As this modest lot has fallen to me, I will return to the pond, notwithstanding the wise admonitions and the bitter tears which I once owed to it. I will return to the pond, but not to that of the small ducks, the pond aflower with illusions: those ponds do not occur twice in a lifetime. For luck like that, you must be in all the new glory of your first breeches and your first ideas.

Many another have I come upon since that distant time, ponds very much richer and, moreover, explored with the ripened eye of experience. Enthusiastically I searched them with the net, stirred up their mud, ransacked their trailing weeds. None in my memories comes up to the first, magnified in its delights and mortifications by the marvellous perspective of the years.

Nor would any of them suit my plans of to-day. Their world is too vast. I should lose myself in their immensities, where life swarms freely in the sun. Like the ocean, they are infinite in their fruitfulness. And then any assiduous watching, undisturbed by passers-by, is an impossibility on the public way. What I want is a pond on an extremely reduced scale, sparingly stocked in my own fashion, an artificial pond standing permanently on my study-table.

A louis has been overlooked in a corner of the drawer. I can spend it without seriously jeopardizing the domestic balance. Let me make this gift to Science, who, I fear, will be none too much obliged to me. A gorgeous equipment may be all very well for laboratories wherein the cells and fibers of the dead are consulted at great expense; but such magnificence is of doubtful utility when we have to study the actions of the living. It is the humble makeshift, of no value, that stumbles on the secrets of life.

What did the best results of my studies of instinct cost me? Nothing but time and, above all, patience. My extravagant expenditure of twenty francs, therefore, will be a risky speculation if devoted to the purchase of an apparatus of study. It will bring me in nothing in the way of fresh views, of that I am convinced. However, let us try.

The blacksmith makes me the framework of a cage out of a few iron rods. The joiner, who is also a glazier on occasion – for, in my village, you have to be a Jack-of-all-trades if you would make both ends meet – sets the framework on a wooden base and supplies it with a movable board as a lid; he fixes thick panes of glass in the four sides. Behold the apparatus, complete, with a bottom of tarred sheet-iron and a trap to let the water out.

The makers express themselves satisfied with their work, a singular novelty in their respective shops, where many an inquisitive caller has wondered what use I intend to make of my little glass trough. The thing creates a certain stir. Some insist that it is meant to hold my supplies of oil and to take the place of the receptacle in general use in our parts, the urn dug out of a block of stone. What would those utilitarians have thought of my crazy mind, had they known that my costly gear would merely serve to let me watch some wretched animals kicking about in the water!

Smith and glazier are content with their work. I myself am pleased. For all its rustic air, the apparatus does not lack elegance. It looks very well, standing on a little table in front of a window visited by the sun for the greater part of the day. Its holding capacity is some ten or eleven gallons. What shall we call it? An aquarium? No, that would be too pretentious and would, very unjustly, suggest the aquatic toy filled with rock-work, water-falls and gold-fish beloved of the dwellers in Suburbia. Let us preserve the gravity of serious things and not treat my learned trough as though it were a drawing-room futility. We will call it the glass pond.

I furnish it with a heap of those limy incrustations wherewith certain springs in the neighborhood cover the dead clump of rushes. It is light, full of holes and gives a faint suggestion of a coral-reef. Moreover, it is covered with a short, green, velvety moss, a downy sward of infinitesimal pond-weed. I count on this modest vegetation to keep the water in a rea-sonably wholesome state, without driving me to frequent renewals which would disturb the work of my colonies. Sanitation and quiet are the first conditions of success. Now the stocked pond will not be long in filling it-self with gases unfit to breathe, with putrid effluvia and other animal refuse; it will become a sink in which life will have killed life. Those dregs must disappear as soon as they are formed, most be burnt and purified; and from their oxidized ruins there must even rise a perfect life-giving gas, so that the water may retain an unchangeable store of the breathable ele-ment. The plant effects this purification in its sewage-farm of green cells.

When the sun beats upon the glass pond, the work of the water-weeds is a sight to behold. The green-carpeted reef is lit up with an infinity of scintillating points and assumes the appearance of a fairy-lawn of velvet, studded with thousands of diamond pin's-heads. From this exquisite jewelry pearls break loose continuously and are at once replaced by others in the generating casket; slowly they rise, like tiny globes of light. They spread on every side. It is a constant display of fireworks in the depths of the water.

Chemistry tells us that, thanks to its green matter and the stimulus of the sun's rays, the weeds decompose the carbonic acid gas wherewith the water is impregnated by the breathing of its inhabitants and the corruption of the organic refuse; it retains the carbon, which is wrought into fresh tissues; it exhales the oxygen in tiny bubbles. These partly dissolve in the water and partly reach the surface, where their froth supplies the atmosphere with an excess of breathable gas. The dissolved portion keeps the colonists of the pond alive and causes the unhealthy products to be oxidized and disappear.

Old hand though I be, I take an interest in this trite marvel of a bundle of weeds perpetuating hygienic principles in a stagnant pool; I look with a delighted eye upon the inexhaustible spray of spreading bubbles; I see in imagination the prehistoric times when seaweed, the first-born of plants, produced the first atmosphere for living things to breathe at the time when the silt of the continents was beginning to emerge. What I see before my eyes, between the glass panes of my trough, tells me the story of the planet surrounding itself with pure air.

NOTES

1. The war of 1830 with Algiers. – *Translator's Note.*
2. The Château de Saint-Léons, standing just outside and above the village of Saint-Léons, where the author was born in 1823. – *Translator's Note.*

The Passionate Observer was illustrated by Marlene McLoughlin
and was set into type by Linda Davis at Star Type, Berkeley,
using Monotype Garamond.

This typeface was originally designed by Jean Jannon,
who was born in 1580 and was the first of the great
typographic artists of the European Baroque.

The type was digitized by the
Monotype type foundry for modern use.

MACR

ecathus pell

BEMBELRAE

BEMBECES

CERCERIS

Anthophorae

SALPINGUS

LUNA

MEGA

POLYCOSA

PHILANTHOMAE

Grashopper

POMPILI

Chalico

Grashopper